T0173157

LAPIDARIUM

Hettie Judah is one of Britain's leading writers on art.
She is the chief critic on the daily newspaper the *i*,
a contributing editor to *The Plant*, and writes regularly
for the *Guardian*, *Vogue*, *Frieze* and the *New York Times*.
Her recent books include *How Not to Exclude Artist
Mothers (and other parents)* (Lund Humphries, 2022),
Frida Kahlo (Laurence King, 2020) and *Art London*
(ACC Art Books, 2019). She lives in London.

www.hettiejudah.co.uk

LAPIDARIUM

THE SECRET LIVES
OF STONES

HETTIE JUDAH

JOHN MURRAY

First published in Great Britain in 2022 by John Murray (Publishers)
An Hachette UK company

1

A CIP catalogue record for this title is available from the British
Library

Hardback ISBN 978-1-529-39494-8
eBook ISBN 978-1-529-39496-2

Designed and typeset in Kings Caslon by James Edgar Studio
Illustrations by Nicky Pasterfield
Printed and bound in Great Britain by Bell & Bain

John Murray policy is to use papers that are natural, renewable and
recyclable products and made from wood grown in sustainable forests.
The logging and manufacturing processes are expected to conform to
the environmental regulations of the country of origin.

John Murray (Publishers)
Carmelite House
50 Victoria Embankment
London EC4Y 0DZ

www.johnmurraypress.co.uk

FOR PHOEBE JUDAH
A DAUGHTER OF Y LLETHR
WHO TREASURES STONES AS SHE TREASURES STORIES

I am not granite and should not be taken for it. I am not flint or diamond or any of that great hard stuff. If I am stone, I am some kind of shoddy crumbly stuff like sandstone or serpentine, or maybe schist. Or not even stone but clay, or not even clay but mud. And I wish that those who take me for granite would once in a while treat me like mud.

•

URSULA K. LE GUIN
'Being Taken for Granite'

CONTENTS

STONES
AND STORIES

STONE
TECHNOLOGY

SHAPES
IN STONE

LIVING
STONES

INTRODUCTION

THE HEADS OF TWO VAST, UNBLINKING GORGONS STARE OUT between the stone columns of Istanbul's palatial underground cistern. Built in the sixth century, during the reign of Justinian I, the cistern was part of the great Byzantine emperor's transformation of the city. Fifteen centuries of drip and submersion have bathed these severed heads an otherworldly algal green. They weren't designed to be here, positioned the wrong way up, supporting columns in the damp darkness. Medusa's petrifying stare once held a protective position overlooking a far earlier structure. Lifted from a despoiled pagan site they were repositioned – and robbed of their talismanic power – by the Christian emperor's stonemasons.

Medusa's myth is a story of stone: the only mortal among the three Gorgon sisters, Perseus beheaded her with the adamantine sword Harpe. Her posthumous gaze could petrify all manner of things: Ovid described Perseus gently placing Medusa's head in a soft bed of shoreline leaves while he stooped to wash his hands, and the plants transforming into the first corals.

A snake-haired woman whose gaze could turn living matter to stone seems an apt response to a world punctuated by monoliths, needles, fairy chimneys, stalagmites and boulders all of which look worryingly poised to spring to life. A petrifying gaze is no more implausible than the belief that a solitary diamond might bind two people in eternal love. Or that mountains grow, continents move and treasures push up from the bowels of the Earth. We humans find it easier to think of abstract notions such as eternity than to enter the geological imagination and follow clues in stone that lead back over four and a half billion years to the fiery origins of our planet.

Historically, stories have helped us make sense of the incomprehensible duration of the world. Tales of an ancient flood helped explain why the shells of sea creatures can be seen in rocks on a mountaintop. Vast fossilised bones and teeth exposed in crumbling cliff faces or washed up on the shore offered evidence of giants that once ruled the earth. Sometimes a rock split to expose the body of an ancient monster. The spectacle of stalactites and stalagmites growing in limestone caves revealed the miracle of stony birth: lithic matter forming in the womb of Mother Earth.

As well as evidence of a long and mysterious past, stone has provided the tools of human progress, from the earliest rough projectiles, through cutting and grinding implements, to the rare minerals that power the present-day information age. As human adornment, stones are associated with wealth, and often, by extension, corruption. Invading armies covered themselves with finery in a show of power, and plundered the mineral treasures of a conquered territory: jewels were ransacked from treasuries; gems, jades and marbles extracted from the Earth; and the carved stones of temples and palaces plundered, to take up new positions in structures like Justinian's cistern.

As humans told stories, and conceived ever more elaborate relationships with their environments, bright pebbles and other lithic treasures were selected to accompany bodies into the afterlife. Markers within the landscape furnished details of creation stories: twinned hollows high in the rock were the eyes of a trickster spirit pursuing sisters across hundreds of miles of dusty territory; pinnacles and slits in the living rock were the dormant reproductive organs of titanic mountain deities. As pigment paint, or hand-held totems, stone was used to express veneration for the spirits that animated mysterious, mutable phenomena: the moon and the tides, winds and weather, abundance and fertility. Monoliths marked out sacred sites and provided a framework for rituals.

We have turned to the beauty of stone to glorify our gods – and ourselves – slicing it to decorate the surfaces of temples and palaces. Michelangelo walked among the stonecutters at Carrara and Pietrasanta, and climbed mountains searching for the sugar-white marble that would yield his sculptures. Beauty ideals of earlier generations are passed down to us in carved stone: alabaster taught eighteenth-century Europeans what human flesh should look like. In the early twentieth century, sculptors allowed stone to lead the way, searching for the forms hidden within.

The alchemists' quest for the mythic philosopher's stone propelled scientific discoveries in India, China, North Africa, the Middle East and Europe. They imagined this stone to be lively stuff, the product of mineral reproduction, able to transform metals and extend human life. Indeed, rock is seldom rock-steady: stones are not still and unchanging, but their transformation occurs on a different – inhuman – timescale. The philosophers of today invite us to think like stone, to abandon our conviction that human life is the measure of all things, and to imagine change instead at the slow pace of a mountain.

Lapidarium is a chamber of stones – a jumbled collection of lithic curiosities. The green copper roof of the lapidarium at the Czech National Museum in Prague shelters architectural fragments and statuary dating back to the eleventh century. The lapidarium at Avignon houses Greek, Roman and Etruscan sculptures, and the chipped remains of early Egyptian and Celtic endeavour. Rather than a museum, this lapidarium takes the form of a storybook. Each tale is led by a different stone, from Alunite to Turquoise, spanning the globe (and beyond), reaching back through history to early Earth. Like waves plucking at shingle, these stories dart from subject to subject, snatching at tumbling fragments from archaeology, geology, mythology, literature, science, sociology and philosophy.

No writer could be expert in so much: coming from art history, I have approached my sibling subjects as a wide-eyed enthusiast and hope these stories will be read in that spirit. Of course there is space for art in any museum, and it crops up often enough in *Lapidarium*, though not always where you might expect it. This is not an art book, but it does explore ideas of great interest to artists: ecology, deep time, how ideas of beauty develop, the rights of natural phenomena, the construction of identity, ancient wisdom, and the abiding sense that certain objects – a stone effigy of Medusa, perhaps – possess an animating spirit.

There are many routes through *Lapidarium*. The stories have been loosely clustered around six themes, but play with them – stones are good for games (try Ducks and Drakes, or Go) – and read them in whichever order you wish. Together, they explore how human culture has formed stone, and the roles stone has played in forming human culture.

STONES
AND POWER

ALUNITE

AMBER

BLACK SHALE

EMERALD

MALACHITE

MARBLE

NEPHRITE

OLD RED
SANDSTONE

RUBY

SAPPHIRE

STONES
AND POWER

THE STONES OF STATE SHINE FROM CROWNS AND SCEPTRES, RAISE and line the great halls. They describe not only the mineral wealth of a territory, but also the reach of its power. In the stones of a palace and its regalia can be read the command of trade routes and control of distant territories: lapis lazuli from Afghanistan, rubies from Burma, Colombian emeralds, or marble from the Mediterranean. Catherine the Great wore the wealth of Russia in jewels stitched onto her bodices. Queen Elizabeth I's robes coruscated with virginal pearls. Gemstones help construct an otherworldly aura of power, like celestial light.

Through legend, stones came not only to express power, but also to bestow it. The Lia Fáil marking the ancient seat of the Kings of Ireland in County Meath was a coronation stone, said to roar when touched by the rightful king. St Edward's Sapphire in the British Crown Jewels was supposedly worn by Edward the Confessor. The godly monarch found himself without alms for a beggar so gave the ring from his finger. Years later, two pilgrims stranded in the Holy Land were offered shelter: their host produced the sapphire ring with a message from John the Evangelist that the King would join him in heaven. The stone was thus considered to endow divine authority.

The struggle for territorial power is often the struggle for mineral wealth: ore, fuel, construction material and other precious substances extracted from the Earth, enriching monarchs (and corporations) hundreds of miles distant. The science of geology does not play a neutral role: there is power bound up in the acts of analysing, categorising and naming things. In the nineteenth century, geological surveys made the race to extract resources more efficient, and provided fuel and materials for expanding empires: the East India Company's 1851 Geological Survey of India identified coal and iron ore to supply the railways.[i]

In *Invisible Cities* (1972), Italo Calvino describes the divided city of Sophronia. On one side are circus acts and rollercoasters, on the other, stone and marble-clad banks, palaces and schools. Half the city is permanent, the other itinerant: 'And so every year the day comes when the workmen remove the marble pediments, lower the stone walls, the cement pylons, take down the Ministry, the monument, the docks, the petroleum refinery, the hospital, load them on trailers, to follow from stand to stand their annual itinerary.'

Stone and marble communicate permanence, and with that, trustworthiness and authority. In London, the architecture of power is dressed in pale oolitic limestone quarried on the Isle of Portland in Dorset. Portland is the stone of the Palace of Westminster, St Paul's Cathedral, the Bank of England, the British Museum, parts of Buckingham Palace and Tower Bridge.

Yet stone ruins are, in themselves, a potent symbol of the impermanence of power: the empire fallen, the despot toppled, the rubble of a plantation house watched over by its ghosts.

#01

ALUNITE

Y OU COULD MAKE A FORTUNE FROM ROCK AND OLD URINE. YOU
just needed the right rock. And the right recipe.

Alum was an alchemist's mainstay. Known and used in China,
North Africa and the Middle East for as long as humans have
brewed potions, this di-sulphate salt is one of the few identifiable
compounds mentioned in ancient proto-scientific texts. It was used
in Mesopotamia and ancient Egypt for tanning, textile production
and (distinctly toxic) medical preparations. Among other things
it acts as a mordant (from the Latin *mordere* – to bite) allowing dye
to take and hold fast. This quality made alum and the recipe to
extract it from alunite stone extraordinarily valuable, and the European alum trade between 1400 and 1700 a mess of wrangling, skul-
duggery and rank manipulation (much of that last conducted in the
name of the pope).

For centuries Europe looked to Constantinople for coloured
textiles. The great domed Byzantine city was the source of silks that
floated on the air and caught the eye like exotic plumage; its work-
shops produced shimmering embroidery, and its merchants supplied
the dyestuffs and mordants that allowed woollen textiles to be
processed in glorious colours. Europe produced woad for blue,
madder for pink and other vegetable dyes, but none of these
matched the intensity achieved with imported indigo, kermes[i] and
saffron. In return for silks, dyes and spices were traded timber,
honey, salt, wax, furs and, until the ninth century, enslaved
members of other European tribes.[ii]

In 1453 Constantinople fell to the Ottoman army. The great
Muslim empire led by Sultan Mehmet II supplanted Christian
Byzantium and gained control of trade routes through the Eastern
Mediterranean. Italian merchants traded with the Ottomans for
alum and dyestuffs, but it rankled that their textile industry was
held ransom to a hostile and capricious power.

Among the Italians to have worked in Byzantine Constantinople
was Giovanni de Castro, a well-connected textile agent who had
run a dye factory, and seen rock worked for alum near Smyrna.[iii]

De Castro investigated the country around Rome, and found alunite in the Tolfa Hills near Civitavecchia, seventy kilometres from the Vatican. Envoys sent from Rome 'shed tears of joy, kneeling down three times, worshipped God and praised His kindness in conferring such a gift on their age.' Alum works were established, and in production by 1459.[iv] Pope Pius II 'determined to employ the gift of God to His glory in the Turkish War and exhorted all Christians henceforth to buy alum only from him, and not from the Unbelievers.'[v] The European alum trade thus became a papal monopoly, run by the Medici family who engaged muscular tactics to suppress rival sources.[vi]

So began an elaborate dance between the powers of prejudice, fashion, economics and the Catholic Church. The vivid textiles with which Byzantium once led European fashions now acquired the 'taint' (as in tinted – dyed – but also morally blemished) of the East. Bright colour was non-European. The great European courts instead turned to black, led by Charles V of Spain, who, by no coincidence at all, also had dominion over Flanders – since the eleventh century, the centre of the north European textile industry and source of luxurious, saturated black cloth.[vii] Importing wool from England, they dyed first with woad, then in madder and alum. Flanders supplied the Catholic courts of Charles V and Philip II with their gorgeously worked black garb. Philip II would only receive visitors to court if dressed in black, which in turn enriched the northern parts of the empire. Dress codes set by the royal court stimulated an appetite that trickled down through all social ranks.[viii]

Later, black clothing would be adopted by the godly Protestants of Northern Europe, for it answered both to a love of fine things and desire for outward modesty: monochrome set it apart from imported luxuries. As demand for alum rose, so did the price set by the Vatican.

Born Giovanni di Lorenzo de' Medici, Pope Leo X spent his own fortune and that of the Vatican on art, books and the construction of St Peter's Basilica. To recoup funds he raised the price of alum and ordered the selling of indulgences – the waiving of penance for sins and wrongdoing in return for a fee. As a consequence of the former policy, Flanders and other territories succumbed to the temptation of trade with the Ottomans. As a consequence of the latter, the German theologian Martin Luther made public his

'Disputation on the Power and Efficacy of Indulgences', starting a chain of events that would lead to his excommunication and the Protestant Reformation.

As the Flemish had before them, the English realised that selling undyed wool was less profitable than trading finished cloth, and from the early sixteenth century took measures to improve their domestic textile industry. In 1545 the now Protestant Henry VIII made a substantial purchase of alum from the papal works, for which he paid with the large quantity of lead mysteriously at his disposal after the dissolution of England's monasteries.[ix]

There would not always be lead to trade, but alum was needed if England was to produce finished cloth. Monopolies for the search and production of English alum were granted as early as the reign of Elizabeth I, though with no success. Prolonged hostilities with the Spanish had seen boats laden with alum apprehended as they passed through the straits of Gibraltar. The state of the domestic dyeing industry was parlous, even by contemporary accounts: 'Anyone who can afford it wyll not meddle with any cloth that is dyed within this realm.' Dyers' lists of the late sixteenth century reveal such appealing shades as sheep's colour, motley, new sad colour, puke (a blue-black), devil in the hedge (off-red), pease-porridge tawny and goose-turd green.[x]

In 1607, alum shale was finally discovered, by a consortium led by Sir Thomas Chaloner, in north-east Yorkshire. Chaloner was a worldly man, and had travelled to mines in Germany and Italy and visited parts of the alum works at Civitavecchia. He had found the rock, but lacked the recipe. Unable to break into the private workings of the alum house, he went one better and persuaded Italian workmen to accompany him back to England. According to the second-bottle version of the tale, Chaloner had two workmen smuggled out of the port of Civitavecchia hidden in a barrel.[xi]

Whitby became the centre of English alum production, though it took decades to manufacture the compound in profitable quantity. Gouged out of the cliff face, and dropped onto pyres of brushwood, the carbon-rich alum shale was roasted at a slow burn for nine months. By then, the rock had turned powdery, a mix of ferrous sulphate and aluminium sulphates mingled with insoluble silicate residue. The substance was then washed through with water to capture the sulphates, and the liquid carried to the alum house to be converted into ammonium aluminium sulphate by the addition of urine, shipped to Whitby from Sunderland and Newcastle. The final step was to bring this delectable potion to the precise point of saturation at which alum would crystallise leaving behind unwanted ferrous sulphate. This, writes geological historian Roger Osborne, was the nub of the alum maker's secret, the moment of alchemy that could turn rock and urine into gold: if an egg placed in the alum liquor floats to the surface, the concentration is right. [xii]

With its own source of alum, England was able to produce finished woollen textiles for international trade. Cloth in sombre tones – once favoured because of the cultural and economic boycott of the 'exotic' East – became the mainstay for gentlemen's suiting. The penguin dress code of business, law and diplomacy has come down to us through the earliest chemical industry, centuries of hostilities between Europe and the Ottoman Empire, papal trade monopolies and great vats of rock and urine.

#02

AMBER

ELEKTRON IS A GREEK NAME FOR AMBER (IT MIGRATED INTO Latin as *electron*). In the third century BCE, the natural philosopher Theophrastus observed amber's static electricity, a 'power of attraction' which he likens to a magnet. He also describes the curious substance lyngourion, which shared amber's powers – indeed it was simply the stone by another name – 'some say that it not only attracts straws and bits of wood, but also copper and iron, if the pieces are thin.' Lyngourion was supposedly formed from lynx urine: 'better when it comes from wild animals rather than tame ones and from males rather than females.'[i] Theophrastus's *On Stones* remained a source for lapidaries until the Renaissance, and the formation of lyngourion a popular fixture for illustration.

William Gilbert was the first to give a name to an 'electrick force' in *De Magnete* (1600), where he proposed that the Earth is a giant magnet. Building on the work of Theophrastus, Gilbert documented the origins of amber and its performance in a series of experiments. Amber 'comes for the most part from the sea, and the rustics collect it on the coast after the more violent storms, with nets and other tackle.' It is the sea water that firms it, thinks Gilbert, 'for it was at first a soft and viscous material' that permitted the entombment of the flies, grubs, gnats and ants now 'shining in eternal sepulchres'.[ii]

Through his exploration of 'electrick force' Gilbert came to understand the nature and limits of amber's static electricity, and its power relative to other substances: 'Amber in a fairly large mass allures, if it is polished; in a smaller mass or less pure it seems not to attract without friction.'

In Gilbert's time the principle source of amber was the German state of Prussia, which controlled the southern Baltic coast. Here the stone had been central to power plays of a political rather than a geological kind. For centuries, the region had been the monastic state of the Teutonic Knights, a bellicose, acquisitive order formed in Palestine in the twelfth century. In 1230, the Teutonic Knights were invited to engage in a northern Crusade to convert the pagans of Old Prussia.[iii] By merciless use of both sword and gibbet, they subdued the Prussians and, as was their wont, took control and expanded across the region. This is prime amber territory and in taking Old Prussia, the Teutonic Knights also took control of the lucrative amber trade, and its routes to Rome, Athens and Constantinople.

Laced with tiny bubbles, most amber sinks in fresh water, but in the brine of the sea it dances, half suspended in the waves, when loosened from its ancient subaquatic bed. Until the nineteenth century, sea-borne amber was so plentiful that no one thought to mine it: why bother when the stones arrived in the shallows, or washed onto the beach? The Teutonic Knights entered the gem trade with all the charm and equanimity they had brought to baptising Old Prussians. They set up strict rules for collecting, carving and dealing: apprehended smugglers were hanged from the nearest tree.[iv] The destiny of this brutally controlled material was rosary beads, devotional images and carved saints. Trade was brisk, and profitable, 'for throughout Christendom no price was too high for a rosary strung with lucent amber beads'.[v]

Once the healing, antiseptic sap of ancestral pine trees, amber is petrified but retains its chemical composition. To qualify as amber, tree resin needs to polymerise, a transformation that can take millions of years – the oldest retrieved amber dates back to the Upper Carboniferous, about 320 million years ago. Colours range from blood red to a milky white caused by tiny air bubbles: evocatively named varieties include Fatty Amber, Foamy Amber and Cloudy Bastard Amber. Dabbed with alcohol or acetone, true amber will not become sticky: if it's susceptible to solvents, the substance you're holding is merely hardened resin (copal). As its German name suggests – *Bernstein* – it takes a flame keenly. Rubbed, amber releases a resinous aroma, a sought-after quality for hand-held amulets and rosaries.

For those in the business of graven images and devotional accessories, the Protestant Reformation in the sixteenth century was disastrous, but by then the Teutonic Knights' powers were already weakening. In 1525, under the influence of Martin Luther, the Knights' Grand Master Albrecht converted the monastic state into a secular principality, became a Lutheran, was granted the title duke of Prussia, and married a Danish princess. The remaining

Knights either followed him into respectable matrimony or left for Rome.

The amber was still there, washing onto the beach, the legacy of long-since migrated ancient forests. The Teutonic Knights left a legacy as well. In 1699, Frederick III, elector of Brandenburg, duke (and, later, first king) of Prussia discovered a huge cache of amber in Königsberg Castle, former seat of the Grand Masters. He commissioned the sculptor Andreas Schlüter and a Danish master-craftsman to construct him a chamber lined in panels of the rich honey-coloured stone, which would glow like summer sun in the Baltic winter. The amber was treated, warmed, oiled, manipulated, moulded and set into panels, though the craftsmen never worked out a satisfactory means of securing mosaic to wood. There were arguments, sackings and incarcerations: the chamber was never completed, and was placed in storage after the king's death.[vi]

Frederick's son Frederick William I, known as the *Soldatenkönig* (soldier king), was less interested in glowing amber chambers than in amassing (and breeding) a regiment of outsized infantrymen. Officially known as the *Potsdamer Riesengarde* (Potsdamer giants), the minimum height requirement was 1.88 metres, and many of the king's prized recruits far exceeded that: locally they were known as the *langen Kerls* – long blokes.[vii] Seeking an advantageous alliance with Russia, the giant-obsessed king sent his father's theoretically magnificent (if in fact unfinished and fragmenting) amber chamber to Peter the Great. The tsar was initially delighted, and sent Frederick William a clutch of outsized soldiers as a thank-you gift. He was less thrilled when the parlous state of the chamber was revealed.[viii]

In 1755 the chamber was completed by Peter the Great's daughter, the Empress Elizabeth, as part of her extravagant design for the summer palace at Tsarskoye Selo, thirty kilometres south of St Petersburg. It took many years and vast expense, but Schlüter's panels were finally installed as the much expanded Amber Room: a glowing, jewelled chamber, lined in stones more precious than gold. Dismantled and shipped back to Königsberg by the Nazis during the Second World War, the room has since disappeared: one of the world's great missing masterpieces.

The philosopher Immanuel Kant, a resident of Königsberg, looking within the amber to the insects, pollen particles and tiny lizards, saw a mute connection to prehistory: 'If thou couldst but speak, little fly, how much more would we know about the past.'[ix] Two hundred years later, a similar – if perhaps less elegantly expressed – thought would inspire the film *Jurassic Park* (1993).

Observing lodestone's ability to attract iron, and amber's to attract straw, the pre-Socratic philosopher Thales of Miletus was said to have suggested that lifeless things might also have souls.[x] In our time of artificial intelligences, and the legal 'personhood' of rivers and mountains, it is an ancient thought that feels surprisingly modern.

#03

BLACK
SHALE

RACKING — HYDRAULIC FRACTURING — EXTRACTS GAS AND OIL
from deep shale formations. The shale is unyielding, and
requires energetic persuasion. A well is drilled down over two kilometres, through the water table and layers of sedimentary rock.
Once it hits the shale, the drilling is extended horizontally, along
the formation of gas- or oil-bearing rocks. Huge quantities of water,
laced with sand and chemicals, are forced down the well, cracking
the shale walls around the horizontal shaft, which are thus forced
to yield the up-to-now tightly packed fuel.[i] As one engineer in
North Dakota put it: 'Pumping fractures rock. Fluid invades fracks.
Oil comes to Papa.'[ii]

As an extracting technique, fracking has been used by the oil and
gas industries since the 1940s. Horizontal drilling, which requires
great technical precision to follow the thin, undulating seam of
black shale for kilometres beneath the surface of the Earth, has only
been in practice since the late 1990s. The process guzzles water:
on average twenty million litres are forced into each well at high
pressure. Mixed into that are 200,000 litres of acids, biocides,
scale inhibitors, friction reducers and surfactants: a cocktail of 750
chemicals the precise composition of which is kept secret in
the US thanks to the 2005 'Halliburton loophole'.[iii] The sand
accompanying this toxic liquor is forced into the emerging cracks,
propping them open. As the water comes back up the well, it also
brings heavy metals, hydrocarbons and radioactive materials from
the shale itself.

Fracking is controversial for many reasons. It has been connected
to minor earthquakes, and uses and contaminates huge quantities
of water in drought-prone areas. In Pennsylvania drillers were
permitted to discharge much of their waste through sewage
treatment plants. An investigation by the *New York Times* in 2011
found that six billion litres of waste water had been produced by
Pennsylvanian wells over three years, and that most was sent to
sewage processing plants unequipped to remove radioactive and
other chemical contaminants before the water was discharged into

lakes and rivers.

The writer and land use activist Lucy Lippard has dubbed these acts of violent, deep extraction 'undermining'. Undermining in a literal sense, in that it creates pits and shafts that alter irreplaceable ecosystems. Also in a symbolic sense, in that profit and undiminished energy consumption are placed ahead of human health, fragile ecosystems and sacred landscapes. Lippard describes the long-standing 'Split Estate' rule by which a landowner in the US can sell land twice: the territory of the surface, and the mineral rights beneath it. The recent uptick in fracking has put a new complexion on Split Estates: 'the practice of horizontal drilling from next door, or miles away, to get at thin layers of gas is especially frightening.'[iv]

As long as oil and gas prices remain high, fracking is big bucks, and a big source of employment: powerful inducements for a state to issue licences. Such clout has made it tough for residents and environmental groups trying to connect water contamination to fracking activity.

In the small town of Pavillion, Wyoming, degradation in water quality coincided with an upswing in gas development during the 1990s and 2000s. In 2008, residents complained of a foul taste and odour, reported methane coming from their taps, and started to question whether their water was making them ill.

Over a decade of controversy followed, the pendulum swinging back and forth between different interest groups. In 2011 a draft report from the Environmental Protection Agency (EPA) found benzene and other fracking-related chemicals in a deep freshwater aquifer.[v] The oil and gas industry pushed back hard and, after three years, the EPA handed the investigation over to the state, which in 2016 announced that fracking was not the cause. That same year, a Stanford University study reached the opposite conclusion, swinging the pendulum back against the energy companies.[vi] A 2019 report from the Wyoming Department of Environmental Quality (part funded by gas company Encana) cleared the frackers of blame.[vii] Meanwhile the people of Pavillion still live with dangerously contaminated water.

As with all fossil fuels, shale oil and gas releases captured carbon into the atmosphere, contributing to the greenhouse effect which in turn boosts global warming. The black in black shale is carbon. These are mud rocks with a high proportion of organic matter: dead stuff, deposited at the bottoms of ancient oceans and lakes. It is widespread: on the map, the coloured patches of shale oil and gas deposits are spread around the world, with large fields in Russia, China and the Americas.

Black shales have been deposited in episodes known as Oceanic Anoxic Events: times when the deep sea has been severely depleted of oxygen.[viii] They are closely tied to episodes of mass global extinction. The most recent mass extinction came at the end of the Cretaceous, sixty-six million years ago, and did for the non-avian dinosaurs. (Cretaceous black shales are a major source of petroleum.)[ix] This was preceded by four other episodes: end-Ordovician 440 million years ago; late Devonian 372 million years ago; end-Permian (aka the 'Great Dying') 252 million years ago and end-Triassic 201 million years ago.

The Bakken shale formation underlying parts of Montana, North Dakota, Saskatchewan and Manitoba dates from the end of the Devonian period – this was the oil field that led the fracking 'revolution' in the 2000s. Mass extinction sounds misleadingly rapid: instead it happened in pulses over the course of 25 million years, starting about 380 million years ago. The Devonian was the age of the fish (vertebrates had just made their first ungainly slither onto land) with the oceans ruled by heavily armoured placoderms, the greatest of which was dunkleosteus, nine metres long with a helmet like an executioner and a bite force like a Great White.[x]

When the oceans became anoxic, corals, sponges, trilobites and other smaller creatures as well as dunkleosteus and her cousins would have been asphyxiated. What caused the anoxia remains a mystery: perhaps volcanic activity, an asteroid, massive algal blooms causing ocean dead zones, ultraviolet radiation penetrating a hole in the ozone layer, or climate change caused by the proliferation of the first land plants. It was probably a combination of the above. Three-quarters of all plant and animal species were wiped out, their digested remains ultimately making their way to the anoxic ocean depths, preserved and building up slowly into the muddy layers that would eventually compact into black shale (see also: Slate, p. 291).

'While we may have heard too often that "everything is connected," there is no longer any doubt that it is,' writes Lucy Lippard. 'We ignore this truism at our risk. Water drained from the land undermines everything.'[xi] There is a grotesque circularity to the fracking of black shale, and the water-use, chemical contamination and global warming that its energy brings. Mass extinctions furnished the fossil fuels through which humankind is accelerating its own extinction, and that of the species with which it shares the planet.

#04

EMERALD

THE MAHARANI OF KAPURTHALA'S FAVOURITE JEWEL WAS AN emerald shaped like the crescent moon, which, as an eighteen-year-old newlywed, she had spotted shining from the forehead of her husband's oldest elephant. 'I thought it was a pity that such a beautiful stone should be worn by an elephant, and I asked the Maharajah for it,'[i] she later wrote. He objected that the gem was too big and coarse for her, but she was not to be denied. Portraits of this legendary beauty show her wearing the enormous stone just as the elephant had, on her forehead. If that sounds like an episode from a fairy tale, that is much in keeping with the maharani's own story, which had played out according to the ruinous fantasy peddled to penniless girls: that if you are beautiful, good and sharp witted, you may one day marry a prince.

The Maharani Prem Kaur of Kapurthala had been born Anita Delgado Briones in Málaga in the south of Spain in 1890. Her parents ran a café and gambling den, but when the government cracked down on illegal gaming, they moved to Madrid seeking alternative employment.[ii] In the capital, their two daughters exploited their modest dancing talents and far from modest beauty, performing as 'Las Hermanas Camelias' – the Camellia Sisters – a support act at the splendid Gran Kursaal cabaret.[iii] A photograph shows the sisters in lacy shirts with camellias pinned in their thick dark hair.

In May of 1906, crowned heads gathered in Madrid for the wedding of King Alfonso XIII and Princess Victoria Eugenie of Battenberg. A modish haunt for artists and writers, the Gran Kursaal attracted aristocratic visitors in town for the celebrations, among them the immensely rich Maharaja Jagatjit Singh of Kapurthala, who as per fairy-tale convention, fell in love with Anita Delgado at first sight. The intellectuals and artists who frequented the Gran Kursaal decided that Anita should marry the maharaja and made it their mission to bring the pair together.[iv]

The Spanish royal wedding ended in catastrophe when the revolutionary Mateo Morral threw a bomb concealed in a bouquet at the royal carriage, killing dozens of bystanders. Festivities suddenly over and Madrid in uproar, Jagatjit Singh fled for Paris. After a flurry of impassioned correspondence from the maharaja (and a large placatory payment to her parents) Anita was sent for. She and her family made the journey to Paris on the understanding that she would accept his hand in marriage. The first task was for the sixteen-year-old dancer to learn French, so that she could understand the maharaja's expression of love and subsequent proposal. Thereafter, Anita submitted to the full Pygmalion treatment, acquiring the accomplishments, manners and dress befitting a maharani. She learned to dance, skate, ride a horse, play tennis, billiards and the piano, acquired several languages and was rigorously schooled in etiquette. In January 1908, appropriately transformed, she travelled to Kapurthala and became Jagatjit Singh's fifth and (supposedly) favourite wife.

The francophone Jagatjit Singh maintained a grand pavilion in Paris, modelled his palace in Kapurthala on the Château de Fontainebleau and was rumoured to drink only Evian water. An important client for Cartier, he was one of the first maharajas to bring jewels from his family's collection to be reset in the contemporary European style. Many of these treasures were worn by the elegant maharajah himself. In preparation for his golden jubilee in 1926, Cartier re-designed the Kapurthala headdress, a tiara carrying nineteen emeralds set in gold, of which the largest weighed 177.4 carats.[v] When the German navy sunk the S.S. *Persia* off the coast of Crete in 1915 it took with it gems belonging to the maharaja now thought to be worth $50 million.

Jagatjit Singh's new maharani, now known as Prem Kaur, fairly dripped with jewels. The coveted crescent-shaped emerald was presented on her nineteenth birthday as a reward for learning Urdu. Her husband told her she now owned the moon, but doubted she could wear it. He challenged her to display the unwieldy gem at the party that evening. 'I was left alone,' she wrote, 'and looked carefully at the magnificent stone, which was huge and with the edges set in a very fine ring of gold. I noticed that in the corners there were two small holes; with great care, I managed to slip a golden thread between the setting and the stone, at the height of the two peaks of the moon. With the thread hidden within my combed hair, the emerald shone on my forehead.'

That night, she was dressed in a green sari, 'matching the colour of my crescent, and I appeared at the party wearing my husband's gift'.[vi] On her next trip to Paris, she had the emerald set in platinum surrounded by diamonds in the Belle Epoque style: in official portraits she poses like a starlet in an embroidered sari with the jewel at her forehead.

Emeralds were rare in the old world. Ancient treasures were drawn from Cleopatra's mines, but other gems had only surfaced occasionally at sites in Austria, Afghanistan and Pakistan. This changed when Hernán Cortés arrived in the Americas. The first jewels sent back to Spain by the conquistadors were simply plundered – first from Aztec temples, and then from the Incas, who worshipped emeralds and believed their light was divine. In the mid-sixteenth century the Spanish fought their way towards sources of emeralds in the Cordillera Oriental in what is now Colombia. There they enslaved the local people and took brutal control of extraction.

The mines produced emeralds of exceptional quality and size. Pirates in the Caribbean really were chasing a Spanish fleet laden with treasure. The missionary José de Acosta wrote that the boat on which he returned to Spain carried two chests of emeralds, together weighing 100 kilograms. Six thousand uncut stones were recovered from a wreck that went down off the coast of Florida in 1622.[vii]

So many emeralds poured out of Colombia that by the early seventeenth century, their value plummeted: searching for a new market, the Spanish turned to India, which had built its magnificent wealth on trade in spice, textiles and gemstones. Until the eighteenth century, India had the world's only source of diamonds, as well as supplies of gold, but it didn't have emeralds. The Portuguese controlled the shipping route from Europe to India, and over the course of the year-long trip to Goa, the emeralds they carried often acquired an alternative provenance: the Mughals were suspicious of the new mines, so the finest of the Colombian emeralds were re-branded old 'oriental' gems, supposedly from Egypt.[viii]

The maharani's fairy tale ended in 1925, when the maharaja moved on to his sixth wife, the Czech-born Eugenie Grosupova. Prem Kaur moved to Paris with her secretary and her magnificent jewels, living a life of considerable discretion in return for an equally considerable allowance. At her death in 1962, her jewels passed to her only son, Ajit Singh. When the crescent-shaped jewel that had once sat on her forehead was examined in preparation for sale at Christie's in 2019, it was identified as one of the great emeralds from Colombia.[ix] Plucked as a gift from an elephant's head in 1910, it was not the first time this emerald had been in Spanish hands. Prem Kaur considered it the lucky charm that had guided her through a storybook life: 'No matter how many jewels I buy or are given to me, the crescent emerald will always be my talisman.'[x]

05

MALACHITE

IN JANUARY 1880, PRINCE PAUL DEMIDOFF ASTONISHED FLORENTINE society by announcing the sale of the Villa San Donato and everything in it. The entire contents of the palace – from ornaments containing the hair and milk teeth of the Bonaparte family, to marble sculptures of the same by Canova – were to be auctioned. The grand public sale would commence mid-March and, in preparation, the prince had the palace repaired and perfected, with his collection arranged as if ready to receive a deputation of crowned heads.

Acid tongues suggested the Villa itself, built on the swampy grounds of an old convent, was unlikely to find an eager buyer. Writing in the *New York Times*, the great Florentine critic and connoisseur James Jackson Jarves dismissed the prince's collection ahead of its public display, condemning it as 'the result of enormous wealth and untrained taste'. In March, for two weeks before the sale, the villa was opened to the public: 100,000 dealers, agents, socialites, collectors and lookie-loos from across Europe trod its marble floors. Eight hundred people travelled for the sale itself, which left the hotels of Florence 'taxed to capacity'.[i]

Jarves rapidly retracted his condemnation of the prince's collection, issuing a public apology in the *Times* ahead of the sale, noting that San Donato 'reigns supreme' in the 'variety, costliness and exquisite taste displayed'.[ii] Among the magnificent works acquired by New York's great families was a monumental malachite vase – almost three metres high, its surface awash with ripples of mineral green and ornamented in gilded bronze – purchased by William H. Vanderbilt.

Malachite is an assertive presence. A minor ore of copper, when sliced it reveals an intense pattern of concentric banding, like eyes or waves. For more than sixty years, the malachite vase was 'the most conspicuous single ornament in the Vanderbilt house at 640 Fifth Avenue'.[iii] Placed first in the marble-clad vestibule of the new mansion, the malachite vase stood its ground despite extensive remodelling undertaken by the next generation. It dominated the entrance hall until Mrs Cornelius Vanderbilt sold it to the Metropolitan Museum of Art in 1945, following the death of her husband and forced departure from the Fifth Avenue mansion. The flashy vase was out of step with contemporary tastes, and even the Met's curator admitted the piece would be divisive: 'Some will admire it, and others will find little to be said for it.'[iv]

As Jarves had taken great delight in pointing out, most of the prince's collection had been acquired in a greedy rush in the decade before the sale. The malachite vase was not of that order: it had been commissioned for San Donato by his great uncle, Count Nicholas Demidoff, in 1819, and was perhaps the most visible link, in the entire inventory, to the family's Russian origins and the source of their extravagant wealth. Family lore has it that an ancestor in the late seventeenth century was a gunsmith of such remarkable skill (and proficiency in finding mineral deposits) that he was ennobled by Peter the Great, given a number of estates – complete with the serfs to work them – mineral-rich land, an ironworks and the right to search for copper in the Ural Mountains.[v]

That ancestor, Nikita Demidoff, first spotted malachite – a telltale green ore of copper – near Nizhne-Tagil'skoye in 1725 when prospecting for the tsar. The Demidoff mines contained some of the most important copper deposits in Imperial Russia and, as tastes developed for the rich green stone, they also became the pre-eminent source of malachite. The counts Demidoff had a hand in popularising its use for lavish interior schemes of the early nineteenth century (see also: Aquamarine, p.170).[vi] In St Petersburg, the family's 1836 mansion was designed by Auguste de Montferrand, moonlighting from his role as architect of St Isaac's Cathedral. The green stone columns and fireplaces of the Great Hall were so magnificent that they supposedly inspired Tsar Nicolas I to commission his own Malachite Room for the Winter Palace.[vii] The Demidoff mines supplied the malachite that clad the columns of St Isaac's and the Malachite Room; the Winter Palace also houses a number of close siblings to the Vanderbilts' vase.

None of these monumental objects is solid malachite: they are veneered in 'Russian Mosaic', thin slices of the stone set in a malachite-powder paste that gives the illusion of a continuous surface.[viii] The technique was achieved by a combination of Italian and Russian skill. Count Nicholas Demidoff arrived in Italy as

Russian ambassador to the Grand Duchy of Tuscany in the early nineteenth century, and commissioned work from Florentine artisans: lapidaries from Carrara were also working St Petersburg.[ix] Throughout the nineteenth century, diplomatic gifts in mosaicked malachite (and there were many) reminded foreign powers of the country's great mineral wealth: the green stone became synonymous with Russia.

Born in the Urals in 1879, Pavel Bazhov knew about Russian mineral wealth at the stony end: his family were former serfs in the mining region. In 1936, during Stalin's Great Terror, Bazhov started to serialise a set of stories that would eventually be published as *The Malachite Casket*. Dark fairy tales, Bazhov presented his stories as local folklore, a prudent piece of fiction in a dangerous time. Set in the nineteenth century and written in Urals dialect, *The Malachite Casket* features brutal overseers, exploited miners, gossipy villagers, a changeling, magic lizards and a mercurial green-clad goddess from the subterranean depths called Mistress of the Copper Mountain.

In the opening story, a miner is enchanted by the Mistress of the Copper Mountain, and finds rich deposits of silken malachite that are sent to St Petersburg for St Isaac's and the Winter Palace. Many years later, the miner's daughter – a beautiful changeling – so bewitches the young master of the region that he takes her from the Urals to St Petersburg where she dons enchanted jewels once owned by the Mistress of the Mountain and demands to see the room at the palace clad in her father's malachite. After refusing the young master's offer of marriage and mocking him and the tsarina, the beautiful peasant girl melts away into the malachite itself, and her jewels turn to tears and blood.

Although the family name is never mentioned, the mines of Bazhov's stories are the Demidoffs' – source of malachite for the great buildings of St Petersburg – and they are the foolish, greedy masters. Rather than an emblem of power and wealth, in Bazhov's tales the hypnotising pattern of malachite is infused with uncontrollable magic, and the stone refuses to be separated from the dark place of its origins.

#06

MARBLE

ONE AUTUMN IN THE 1980S, WRITER JOHN MCPHEE FOLLOWED the celebrated geologist Eldridge Moores – champion of plate tectonics – around sites on the Mediterranean. Stopping off briefly in Athens, the pair found themselves near the Acropolis, the city's ancient raised citadel and home to that radiant symbol of ancient Greece, the marble temple of the Parthenon. Moores takes McPhee on an impromptu tour. McPhee's attention is drawn as he ascends, to a sudden change from red sedimentary rocks to the massive freestanding block of limestone forming the rocky foundation of the Acropolis itself.

As Moores talks, students gather, assuming him to be an English-speaking guide. He describes the nature of limestone, its vulnerability to water and tendency to form caves. Caves had sacred associations – home to gods and spirits – but they were also useful places of refuge with a ready supply of water. This would have been a good situation from which to wait out a siege. Moores speculates that the limestone of the Acropolis is a klippe – a large body of rock broken away from its place of origin – most probably from the Hymettus Mountains fifteen kilometres to the east. He has other theories. Jargon at last gets the better of the trailing students, and Moores half apologises to McPhee: 'I have always thought it sacrilegious to come here and do geology.'[i]

So much of what we think of as culture – our modes and places of worship, the tools we use, the materials in which we adorn ourselves, the stories we spin, our graven images – is formed by geology. Far from being sacrilegious, in Greece geology is the fastest route to the ancient gods. It's not only the Parthenon that sits on limestone – the Aegean abounds in the stuff, and its islands are fretted with the caves, sinkholes and subterranean spectaculars formed by the passage of water through this soluble, stalagmite-forming rock.

These wild grottoes and caverns in the limestone were the first Greek temples: sanctuaries where new stone (stalagmites) could be seen forming in the womb of the great Earth Mother. (The association is not the Greeks' alone. The Egyptian word for mineshaft – *bi* – also means uterus.[ii])

Aegean limestone formed beneath the ancient ocean Tethys, named by the Austrian geologist Edward Suess after the Titan water goddess. Tethys was slowly compressed by the movement of continental plates: all that remains of her now are the Mediterranean and Black seas in Europe, and Caspian and Aral seas in Central Asia.

The remnants of Tethys are still being squeezed, as the continental plate of Africa continues to move north east, colliding with and sliding under Eurasia. It's a slow movement, but the pressure it exerts is enormous. Some sixty-six million years ago the collision of the African and Indian plates with the Eurasian plate to the north rucked up rock to form the European Alps and the Himalayas. The boundary between the African and Eurasian plates now lies just to the south of Italy, causing millennia of volcanic activity: Vesuvius, Etna, Stromboli and their siblings. All this pressure and heat caused rocks to metamorphose, transforming some of that Tethys limestone into marble.

Fine-grained marble sparkles and reflects light. Sliced thin it becomes skin-like, translucent, golden. Where the limestone and sandstone detailing of Europe's medieval churches have been softened and worn away over the centuries, the carved marble of Greece and Rome has retained vivid detail. 'How much, one wonders, would Michelangelo have achieved had he been obliged to use only granite?'[iii]

At the end of the eighth century BCE, urban temples start to appear in Greece, drawing worship away from woodland sanctuaries and caves to within city walls. These early structures were roughly built, but a century later, monumental temples in locally available stone start to appear. On the islands of Naxos and Paros, the locally available stone was brilliant white marble, attention-grabbing stuff. Soon every city with the money to ship it wanted Naxian or Parian marble for their temples.[iv]

Naxos was also home to the first marble roof – translucent sheets that made the temple ceiling sparkle like stars, as though heaven had been brought down to Earth. Light seems to radiate from within it. White marble is not passive, blank: in the Mediterranean sun it glows from crystalline depths, suggesting the divine presence of the gods themselves.[v]

The Parthenon (or, to call her by her full name, the Temple of Athena Parthenos) was constructed in around 490 BCE according to designs by the sculptor Pheidias. She was built in stone from Mount Pentelikon, the first monumental Greek structure entirely constructed from marble, and the most decorated temple of its time. After Xerxes ordered the destruction of Athens in 480 BCE, all of the temples on the Acropolis were rebuilt in Pentelic marble to match the Parthenon, already regarded as a paragon of devotional architecture.

Sociologist Richard Sennett, who has spent a lifetime pondering the relationship between the body and the built environment, sees the foundational values of Athenian democracy carved into the marble of the Parthenon: 'To the ancient Athenian, displaying oneself affirmed one's dignity as a citizen.' Young free men of the city learned how to form and control body and voice in the gymnasium,[vi] the school at which you learned to reveal physical and rhetorical prowess. Both mind and muscle were serving parts of a greater body – the city – through participation in warfare and statehood. 'In ancient Greek, the very words used to express erotic love of another man could be used to express one's attachment to the city. A politician wanted to appear like a lover or a warrior.'[vii]

That glowing temple on the hill, the Parthenon, was of a part with this culture of exposure. Like the oiled, sun-bronzed, healthy body of the gymnasium-primed Athenian, it radiated vitality and self-assurance. These bodies were celebrated in Pheidias's frieze, which once flowed around the top of the temple walls (but now resides, in part, in the British Museum in London). The figures that emerge from the marble are silken, taut, implacable, exposed, perfect muscular specimens – whether god, man or horse. These glorious bodies tested the distance between gods and free men. Women, foreigners, invalids and slaves were excluded from that ideal: as Sennett points out, bundled up in the idea of honourable, exposed nakedness celebrated in the Parthenon friezes, was the sense of shame evoked by lesser bodies.[viii]

Other city-states prospected avidly for their own sources of marble. The famous quarries at Carrara – Roman Luni – were not cut until the mid-first century BCE. Until then, Roman temples were timber and brick structures faced in stucco and whitewashed, or built from marble imported (or stolen) from Greece. It was Julius

Caesar who commenced Rome's glorification in marble from the quarries at Luni, and his son Augustus who saw the plans realised. 'The city, which was not built in a manner suitable to the grandeur of the empire, and was liable to inundations of the Tiber, as well as to fires, was so much improved under [Augustus] that he boasted, not without reason, that he "found it of brick, but left it of marble".'[ix]

Like the Greeks, the Romans found in the luminescence of pale marble – often enhanced with wax – an expression of an ethical ideal.[x] For Augustus, immortalised in marble, as so many Roman leaders would be after him, the intrinsic qualities of the material endowed his carved likeness with associations of honesty and moral integrity.[xi]

As Rome's empire grew, coloured marbles from across Europe, Africa and the Near East rolled into the city. In this gaudy lithic parade, the bedrock of conquered territories was exhibited in the built fabric of the city itself. Such abundance laid marble open to abuse: 'To Romans, exotic columns for houses instead of temples implied luxury, arrogance, and moral laxity.'[xii] The most expensive of all imported marbles came from Phrygia – present day Anatolia – and was laced with seams of imperial purple. Conspicuous display of Phrygian marble became a target, as per Horace in an ode that opens, promisingly: 'I abominate the uninitiated vulgar, and keep them at a distance.'[xiii]

If marble described the Roman Empire in its pomp, it also broadcast its decline. By the fifth century, shortages left edifices unrepaired, and new constructions unable to match the glory of the old. Vulnerable structures were plundered for their marble wealth, transforming viable buildings into ruins. The trappings of Rome's glory, left unguarded, were open to architectural cannibalism as the Empire gorged itself on its own marble body.[xiv]

#07

NEPHRITE

Fu HAO WAS FORMIDABLE: A CONSORT OF THE SHANG KING Wu Ding, she controlled her own estate, bore princes and conducted sacred and military duties commonly reserved for men.[i] Mother of the heirs to the throne, she was not a woman to be messed with: the earliest known female general in Chinese history, Fu Hao led the army on a succession of military campaigns (and had the spoils of war buried with her to prove it). As a leader of sacred rituals, she carried out acts of divination and presided over state sacrifices in the king's name.

Fragments of the remarkable life of this late Bronze Age warrior queen survive in some of the earliest known Chinese script, inscribed into flat animal bones used for divination during the Shang dynasty (c. 1600–1045 BCE). Cow scapula and sometimes the plastrons of turtles were pierced with metal rods, then heated and manipulated until a crack appeared that provided an answer to a life-or-death matter of health, harvest, battle strategy, childbirth or royal succession. Question, answer and date of divination were inscribed on each oracle bone, and through these durable objects, concerns about Fu Hao's military campaigns and pregnancies can be read, over 3,000 years after her death.

Fu Hao was buried apart from the rest of the royal graves in the Shang city of Yinxu – perhaps on her own land. That distance kept her tomb from the attention of grave robbers and it survived intact until its discovery in 1976: the only grave to be confidently linked to a person named in the oracle texts. It is a tomb of modest dimensions by royal standards of the time – a stepped, rectangular pit, four by 5.6 metres at its widest, and 7.5 metres deep – but the status of its occupant is evident from the finery that accompanied her into the afterlife. Fu Hao was buried with six of her dogs, and sixteen of her human subjects, all sacrificed to accompany her as she joined the ranks of the royal ancestors. The tomb also contained 1,600 kilograms of bronze objects – including many weapons engraved with her name (some of which had been used to sacrifice the retainers buried alongside her) – and 755 jade pieces, the largest single collection to be unearthed.[ii]

'Jade' is applied to two different hard, translucent, sometimes green, stones: jadeite and nephrite (see: Jadeite, p. 89). Nephrite is the jade of ancient China – a stone more precious than gold, of exceptional symbolic potency. Nephrite's durability served as a material connection to a person's ancestors, and suggested long life, indomitability and the endurance of personal legacy. Rarely the cucumber green we now expect of jade, these ancient pieces are carved in liquid colours ranging from milk to black tea.

The importance of such ancestral links is evident in the collection of jades buried with Fu Hao. As well as finely worked items from her own time, Fu Hao was accompanied by her collection of Neolithic jades, some already 1,000 years old, perhaps discovered in more ancient tombs and taken as plunder. Many were ritual objects including the fine perforated discs known as *bi*, and engraved tubes with a square outer profile called *cong*. Both are common grave goods – with the finest *bi* positioned near the stomach or chest – but their significance remains a mystery. The *bi* may relate to stargazing or navigation – the circle form was said to represent heaven – or derive from an agricultural implement. In the eighteenth century, the Qianlong emperor decided one such *bi* was a cup holder, and had it engraved with a poem to that effect. Jade *cong* and *bi* have been carved since perhaps the fifth millennium BCE: the precise shaping, perforating and carving of these objects would have been time-consuming, tough, highly skilled work. Jade is hard but brittle, not easily split or cut, and had to be ground with bamboo drills and abrasive crystals of quartz.[iii]

The challenge of carving jade can be seen in the relative refinement of the pieces found with Fu Hao: a flat pendant carved in the silhouette of a turning phoenix (*fenghuang*) is beautifully balanced by an extravagant flowing tail. By contrast the three-dimensional carvings – tigers, elephants, dragons – stay close to the blocky form of the original stone. One exception is a small, seated woman with a tail or sash protruding from her back: the jade is deeply worked, and the figure has perforations in her headdress and gaps between her body and arms. Some see her as Fu Hao herself, but the artist Ai Weiwei considers her a mythological, rather than memorial, object. The figure's fish tail 'would mean she's a god or ghostlike figure', explains Ai. 'In the Shang Dynasty, you often saw depictions of humanlike figures with a dragon's head or a fish's tail. They are images of transformation. The kneeling position is common, but the tail and headdress are unique; they don't repeat in thousands of objects that come later.'[iv]

The practice of carrying jade into the afterlife was elaborated in the Han period (202 BCE–220 CE). Early members of the Liu family, their consorts, and the high-ranking elite were buried in full suits of jade, made from interlinked plaques of stone. Before the body was placed in the tomb, its nine orifices were first sealed with jade plugs.[v] Jade was thought to promote longevity and would thus protect the body from decay.[vi] Once in position, the body was adorned with *bi* discs before being dressed in its jade suit.

The Han was the era, too, of royal burials with great terracotta armies: warriors carrying the spoils of battle, their weapons held ready to protect the deceased. Sinologist Jessica Rawson sees a shift in attitudes to death around this time: one driven perhaps by fear of malevolent spirits. Back in the days of Fu Hao, the tomb was a gathering place, visited by the family who kept in contact with the deceased through offerings, prayers and sacrifices. In the Han period, passages to mountain tombs were blocked and cut off from the world of the living, and the dead were buried as if ready to do battle, accompanied by ceramic armies or outfitted with jade armour and weapons.[vii]

Around one hundred such jade suits have been found, all finely made, sometimes held together with gold wire, and closely fitted to the body. The cost and care involved is testament to the protective power jade was believed to bestow on the wearer. The corruption of the body after death was considered the work of demons or evil spirits. Wildly impractical in life, a bespoke suit of jade plaques might afford requisite supernatural protection for battles with malevolent forces in the afterlife.[viii]

08

OLD RED
SANDSTONE

IN 1658, AFTER CLOSE STUDY OF BOTH SECULAR AND BIBLICAL accounts of the history of the world, James Ussher, archbishop of Armagh and Primate of All Ireland provided a precise date for the age of the Earth. The first day of creation was 23 October 4004 BCE – a Sunday 'for as much as the first day of the world began with the evening of the first day of the week'.[i]

In 1701, Ussher's date was still regarded as authoritative and printed in the English Bible. A period of 5,662 years is precious little time for earthly transformations, and few, in any case, were thought to have taken place: we inhabited the world as created. In the eighteenth century, as natural scientists observed layered strata of rock, intrusions of granite and basalt, and seashells in limestone far from the coast, it became ever trickier to balance the story written in the Earth with the story written in the book of Genesis (see also: Blue Lias, p. 262).

In the late eighteenth century, Abraham Gottlob Werner, a charismatic professor at the Freiberg Academy, Germany, propounded a version of the geological creation story – known as Neptunism – in which the stones of the Earth built up within a vast primordial ocean. The first layers crystallised from the water itself and after that accrued from alluvium. Some ocean (rather than land) life allowed for the presence of shells in the later layers, and the theory was elastic enough to accommodate new discoveries. Werner's Neptunist vision allowed geological formations to be universal, extending across the globe in neat layers like a Baumkuchen. Its compatibility with Genesis allowed it enormous reach and influence.[ii]

Werner's theory was not uncontested. This was an age of upheaval. While rather less bloody than those taking place in America and France, Edinburgh was the site of a revolution of ideas that birthed the modern science of geology.

In 1788 a group of three men – James Hutton, John Playfair and James Hall – sailed around Siccar Point, looking for proof of profound movements in the Earth's crust that argued for a past far,

far longer than that proposed by Ussher, or indeed Werner and the Neptunists. Siccar Point is often described as the geologists' Mecca, though it might be more accurately described as their Mount Sinai. For this rocky expanse, amid the sparkling inlets and gentle cliffs of the Berwickshire coast, is where the fundamental principles were laid down.

What they found at Siccar Point was a formation that became known as Hutton's Unconformity. An underlying layer of Greywacke (a 435-million-year-old sedimentary formation of sandstone and mudstones: 'Wacke' is pronounced as in 'wacky') had been so rucked up in collisions of the tectonic plates that its layers were positioned vertically. It was heavily worn, and showed evidence of millions of years of exposure before it was apparently submerged again, to be covered 370 million years ago by a sedimentary formation of Old Red Sandstone.

Over years spent walking and watching the Scottish countryside, Hutton – a doctor turned farmer turned geologist – had observed land eroding and silt being carried away to the sea. He had studied ancient structures such as Hadrian's Wall, and noted how slowly exposed stone was worn by wind and rain. It seemed to him that the process of erosion, sedimentation and lithification must be continuous, and that the Earth at times must have been prone to huge upheavals – perhaps caused by immense inner heat – which caused rock to move and shift. Hutton read in the face of rocks 'the annals of a former world', but since most were covered in mosses and vegetation, he seldom got the full story. The journey to exposed rocks at Siccar Point was a hunt for proof of his theory.

For Hutton's companions, long enthralled by their friend's ideas, this was a transformative moment. 'On us who saw these phenomena for the first time, the impression made will not easily be forgotten,' wrote Playfair, who later translated Hutton's ideas into readable prose for the general reader. Observing the Old Red Sandstone and Greywacke of Siccar Point, Playfair imagined himself 'carried back in time' to the point when the sandstone had just started to accrue in 'the waters of a superincumbent ocean'. From there his mind travelled back to the formation of the Greywacke itself, and then to the silt of the eroding mountains from which that in turn had formed. 'Revolutions still more remote appeared in the distance of this extraordinary perspective.

The mind seemed to grow giddy by looking so far into the abyss of time.'[iii]

In his vision of a churning cycle of movement, erosion, accretion, rise and fall, all powered by the heat of the Earth, Hutton punched straight through the Biblical timeline to a world in which 'we find no vestige of a beginning, no prospect of an end.' His theory came to be known as uniformitarianism and, in opening up great vistas of time, laid the ground for other transformative ideas, not least evolution. During his voyage on the *Beagle*, Charles Darwin read Charles Lyell's *Principles of Geology* (1830–3), a book propounding and expanding on Hutton's ideas. Thanks to Hutton and Lyell, generations of geology students have spent weeks hunched against enthusiastic Scottish rain, tapping at rocks with their hammers.

The Old Red Sandstone has a starring role in another great Scottish adventure. There's a hammer in this story too: a royal though less popular one. In 1296, the English King Edward I, 'Hammer of the Scots', flexed his power by removing the ancient stone on which Scotland's leaders were inaugurated. Carried from Scone Abbey in Perthshire, hundreds of miles south to Westminster Abbey, it was known as the Stone of Scone, or Stone of Destiny.[iv] At Westminster it was built into the base of a ceremonial wooden throne, on which generations of English and British monarchs have been crowned.

The stone's return was frequently requested, and denied. In 1950, a group of four Scottish students took it upon themselves to remove the 152-kilogram stone from Westminster Abbey on Christmas Day, dropping it and chipping one corner in the process, then drove it back up to Scotland. It reappeared three months later at Arbroath Abbey, site of the 1320 Declaration of Arbroath in which the barons of Scotland invited the pope to recognise Scottish independence. Though it was soon returned to Westminster Abbey, another attempt to steal the stone was made in 1970, before it was finally repatriated on St Andrew's Day 1996, on the understanding that it would return to London every time it was required for a coronation.[v]

Legends grow around the stone like moss. Following the Christmas Day heist, the stone deposited at Arbroath Abbey was popularly believed to be a substitute. The abbot of Scone, too, was long thought to have hoodwinked Edward I with a bogus boulder.

Around 1900, there was debate in the Irish press as to whether the stone was the Lia Fáil that stood at Tara, seat of the High Kings of Ireland, from which derives the ancient name of Innisfail. The Lia Fáil let out a roar when the true king stood upon it: it was a magical object with a convoluted legend that led back via necromancers to ancient Greece and Syria.[vi] In the thirteenth century, William of Rishanger thought the stone to have come from the Holy Land, where it had formed a pillow for the patriarch Jacob as he dreamed of angels ascending and descending the ladder to heaven.

Ah, science, you killer of dreams. In 1996, Scottish men of science – spiritual heirs of Hutton, Playfair and Hall – resolved the question of the stone's origin. Before the Stone of Scone was placed on display at Edinburgh Castle, the newly returned rock had a wash and brush-up, and a sample was taken for analysis. As it turned out, its genesis lay not in the Holy Land, not Greece, nor Syria, nor even Ireland, but in rocks formed during the Lower Devonian period near Perth, known as the Scone Formation: it was a piece of the Old Red Sandstone.[vii]

#09

RUBY

IN 1588 THE SPANISH SHIP *GIRONA* WENT DOWN AT LACADA POINT, near the Giant's Causeway off the coast of Northern Ireland. It lay undisturbed for almost 400 years, until the arrival of intrepid underwater archaeologist Robert Sténuit in 1967. Over 6,000 diving hours in surging, icy water, Sténuit and his team recovered much that you'd expect in the cargo of a warship – ingots, coin and heavy guns.[i] More surprising was a remarkable quantity of jewellery, chief among the treasures being a pendant in the form of a gold salamander, its back studded with pink rubies.

The *Girona* was a galleass: a fast warship travelling under the combined power of oar and sail. It had been part of the great Armada raised by Philip II of Spain against the 'heretics' of Protestant England (and against Elizabeth I's privateers, who ransacked Spanish ships laden with treasures on their return from South America).[ii] From Lisbon, the 130 boats of the Armada approached England from the south east, and the Spanish planned to gather reinforcements in Flanders, then sail up the Thames to London.[iii] The English sent fire ships in among the vessels of the explosive-laden Spanish fleet while it was moored at Calais. The Spanish had to slip anchor and sail north. Battered by foul weather and depleted by running battles with the English fleet, the Armada was instructed to return home by sailing up the east coast of England, over the top of Scotland and Ireland and into the open Atlantic.[iv]

Spanish maps of the time showed the west of Ireland to be almost straight: their pilots were not prepared for the ragged coast of Connaught jutting out into the Atlantic. At least twenty-six ships went down off the Irish coast. Their wrecks still haunt the coastline in place names like Carraignaspana (Spaniards' Rock) and Port na Spanaigh.[v] The *Girona* was one of the last to go down. It had been repaired at Killybeg where it took on the survivors of two other wrecks, among them Don Alonso de Leyva, a young aristocrat and military commander of great charisma (handsome too, according to his portrait by El Greco[vi]).

Stars of Spain's coming generation, Don Alonso and his entourage were probably the owners of the finery discovered by Sténuit. As the curator first responsible for the hoard wrote: 'It may seem surprising that so much fine jewellery should be carried by sailors and soldiers on a warship, but the Renaissance was a jewelled age and the Armada an invasion fleet – the proud officers and noblemen aboard had every intention of looking their very best when, hopefully, they would strut, triumphant, through the streets of vanquished London.'[vii]

Now carrying 1,300 souls – more than double its original crew – the *Girona* turned round in the hope of finding harbour in neutral Scotland, and from there perhaps safe passage through the Low Countries. It was overburdened and in dangerous waters when it went down. The bodies of 260 sailors from the *Girona* were recovered and buried at St Cuthbert's Church near Dunluce Castle. Only nine of the 1,300 aboard survived – they escaped to Scotland with the help of the local chief, Sorley Boy MacDonnell.

To modern eyes the jewelled salamander is enchanting – a cute little thing with webbed toes and wing-shaped gills, like a baby dragon. It is also an extraordinary emblem of Spain's wealth and the global reach of its power at the time of Philip II. Its golden body is from territories then under Spanish control in South America. In form it resembles the axolotl, an amphibian native to Mexico, venerated by the Aztecs. The rubies come from Burma, so this is not a South American jewel, but one inspired by an Aztec emblem: spoils sent by Hernán Cortés to Spain in 1526 included a 'winged lizard' or salamander in gold.[viii] Of the nine original rubies that studded its back, only three remain. They are clear, rosy and pale. Unusually for the period, they are true rubies rather than spinel.

Since at least the eleventh century, the most prized rubies came from the mines in Mogok in the north east of Burma, and they were strictly controlled by the king: no foreigner was permitted to visit or know the location of the mines. In medieval Europe, the splendour of the Burmese royal courts was legendary. The Venetian jeweller Gasparo Balbi visited the court of Nanda-bayin, the Burmese king of Pegu in 1583 – no doubt on the hunt for rubies to bring back to Europe – and was granted an audience. Tradition dictated he must present the king with jewels. He gave him a box of emeralds, in ample supply in Europe thanks to Spain's ruthless extracting from mines in Colombia (see: Emerald, p. 31). Word of Spain's territorial ambition had clearly reached Burma. Nanda-bayin concluded their meeting by asking Balbi 'if that King which last tooke Portugall were as great, and if Venice were warlike. To which I answered, that King Philip that had taken Portugall was the potentest King among the Christians, and that the Venetians were in league with him, but had no feare of any, yet sought friendship with all.'[ix]

Rubies were known in Spain as early as the sixth century. Saint Isidore, archbishop of Seville in the early seventh century, describes ruby (then known as carbuncle) in his encyclopaedic *Etymologies*. The stone appears in the category of 'Fiery gems': 'Of all the fiery gems, the carbuncle (*carbunculus*) holds the principal rank. It is called "carbuncle" because it is fiery, like a coal (*carbo*), and its gleam is not overcome by the night, for it gives so much light in the darkness that it casts its flames up to the eye. There are twelve kinds, but the most outstanding are those that seem to glow, as if giving off fire . . . It occurs in Libya where the Troglodytes live.'[x]

A stone that supposedly glowed like fire, the ruby was an apt partner for a salamander, a mythic creature that could not be consumed by flames. As a status symbol for an ambitious nobleman, the pendant represented the vast span of Spain's power and reputation, from South America to Burma. As a good luck charm for a sailor travelling into battle on a ship laden with explosives, it was a talisman against the very real threat of death by fire, though alas it proved less potent as protection against death by water.

#10

SAPPHIRE

O N 25 FEBRUARY 1986, OVER A MILLION PROTESTORS TOOK TO Manila's streets demanding the removal of Ferdinand and Imelda Marcos. Following the assassination of a political rival and a clumsily rigged election, the People Power Revolution was at the gates of the presidential Malacañang Palace. After thirty-one years in power, including eight of brutal martial law, it was time for the president of the Philippines and his notoriously lavish First Lady to cut and run.

The Marcoses fled by helicopter to Clark Air Base, seventy-seven kilometres north of Manila. The following day, with an entourage of ninety, they boarded two American Air Force planes to Hawai'i.[i] Among their ample personal effects were jewels stashed in a box of nappies. Impounded before it entered the country, the jewellery was valued by the United States Customs Service at between $5 million and $10 million.[ii]

Interviewed by filmmaker Lauren Greenfield decades later, Imelda Marcos brushed off the stashed gems as an afterthought: 'A few days before there was a big reception, I had to wear some jewelry. Just as we were told to get into the helicopter, I saw a box of diapers from one of my grandchildren, so I put this jewelry in this box of diapers.'[iii]

That was some understatement. The US customs inventory of the Hawai'i jewels stretches to four pages. It includes crowns and tiaras, an emerald parure, over fifty pearl necklaces – many only provisionally strung – and 300 loose pearls. Among the most valuable individual items was a sapphire necklace, set with five stones fat as cherries, and seven smaller ones, all the brilliant limpid blue of seawater over white sand. The rare matching sapphires were all from Ceylon – Sri Lanka under British colonial rule. The value suggested by customs officials was $376,990, but that erred on the low side. A receipt found at Malacañan Palace records Imelda's purchase of a sapphire and diamond necklace from Van Cleef & Arpels in Manhattan in 1981 for $2.6 million.[iv]

The necklace made a brief public appearance in 2015, when the government of Benigno Aquino III invited Christie's and Sotheby's to appraise 750 pieces impounded from the Marcos family. The hoard was in three parts: jewels apprehended in Hawai'i; pieces abandoned at the Malacañan Palace; and a haul known as the Roumeliotes collection after the Greek friend of the Marcoses was apprehended boarding a flight to Hong Kong with $12 million worth of jewellery in his luggage. 'A customs official said later that the gems, including an emerald the size of a large grape and diamond-studded bracelets and necklaces, were in a special "royal" style favored by Imelda Marcos,'[v] reported the *New York Times*.

'If you put me in there to value the jewellery and I had no knowledge of where it came from, I would say this feels like a royal collection,' enthused David Warren, director of jewellery at Christie's.[vi] 'Royal' again – the repetition is telling. Imelda – former Miss Manila – delivered movie-star glamour, but, as heads of state, the Marcoses styled themselves as monarchs, not mere politicians. At Malacañan Palace, intercom labels for their bedchambers read 'King's room' and 'Queen's room'.[vii] 'By and large, dictators lack Crown Jewels – that is, objects with a history,' writes political scientist Judith L. Goldstein. 'Since they lack an inheritance of objects with pedigrees, they do what other rich people do – they build their own collections.'[viii]

In Europe sapphires carry specific associations with royalty. The blue of heaven captured in stone, they symbolise purity, holiness and monarchy by divine right. Napoleon was irresistibly drawn to the Talisman of Charlemagne, a ninth-century medallion with two large sapphires enclosing what was purported to be the hair of the Virgin Mary. The military leader of the French Republic considered himself Charlemagne's descendant and boasted of wearing the medallion into battle as the Holy Roman Emperor had 900 years earlier.

Most familiar in blue, sapphire gets fancy in colours ranging from peachy orange to near black. It can display asterism – an inner star – be bi-coloured, tri-coloured and even change tone under artificial light. Any colour you like, so long as it's not red: a red sapphire is a ruby. The oldest, best-known source of sapphire is Sri Lanka. As 'Ceylon', the island came under British control in 1815, and the royal family have since adorned themselves in its blue gems. Since 1831, the British coronation ring has been a sapphire, and the stone is traditional for royal wedding and engagement rings.[ix] In 1981, the year Imelda Marcos purchased her $2.6 million necklace in New York, the world swooned as Prince Charles announced his engagement to nineteen-year-old Lady Diana Spencer with a twelve-carat blue Ceylon sapphire.

Jewellery was essential set dressing in the Marcoses' regal narrative, but a $2.6 million necklace is mere detail in the bigger picture. The total figure they skimmed from the Philippines is between $5 billion and $10 billion: the Marcos presidency has the questionable distinction of holding a Guinness World Record for the 'Greatest robbery of a Government'. Assets, including substantial properties, are still being traced around the world. The Marcoses' corruption was not just criminal, it was deadly: during the period of martial law from 1972 to 1981, 70,000 people were incarcerated, 35,000 were tortured and 3,200 were killed.[x]

Soon after the Marcos family fled, the Presidential Commission of Good Government (PCGG) was set up to track down ill-gotten gains, with the aim of liquidating them and directing the funds into government projects. It was slow going, thanks to the family's use of aliases and associates,[xi] and to their litigiousness. Over two decades later, the administration of Benigno Aquino III laboured to tackle collective amnesia about the regime and just how the Marcos wealth had been amassed.

Since 2012, artist Pio Abad has worked with the PCGG's blessing on projects that bring the historic activities of the Marcos family back into view.[xii] Abad points to how Imelda – widowed since 1989 – has leant into the clownish shoe-obsessive persona constructed by the international press, and allowed it to disguise her as a loveable buffoon, cloaking her intellect and ambition. The acquisitive excesses of the Marcos regime made it dangerously easy to ridicule, while ignoring the political dynasty still under construction. For Abad, this work is personal as well as political: his activist parents were both incarcerated under the Marcos regime.

In June 2016, nine months after the government invited Christie's and Sotheby's to evaluate and photograph the jewels, the political climate switched with the election of the Marcos-backed right-wing populist Rodrigo Duterte. There would be no sale, no further investigation: the jewels would simply disappear from public view. Sensing the end of the road, the PCGG sent photographs taken for the cancelled auction to Abad. He couldn't use the actual pictures – those were still government property – but perhaps he could keep the memory of these objects alive, somehow?

Abad turned to his wife, the British jewellery designer Frances Wadsworth Jones, who pieced together detailed digital models of the prize auction lots with lapidary skill. It was no small task: a pearl diadem once owned by the Russian royal family had 20 million facets. 3D printed in white plastic, the jewels became spectral, like cast snakeskin – robbed of their glamour and sparkle. In 2019 these mnemonic proxy jewels were exhibited as *The Collection of Jane Ryan & William Saunders* – pseudonyms used by the Marcos family for Swiss bank accounts – at the Honolulu Biennial in Hawai'i: a return of sorts for the gems the Marcos family had tried to smuggle into the US in 1986. Beside each piece was the given value, and the relative cost of an infrastructure project: the Romanov diadem could 'provide housing to 1,200 homeless beneficiaries'; a diamond, sapphire and ruby bracelet 'is worth the full immunisation of 20,000 children plus 17,600 pneumococcal vaccines to senior citizens and infants'; the sapphire necklace could connect over 2,000 homes currently without an electricity supply.

'It's a bonny thing. Just see how it glints and sparkles,' said Sherlock Holmes, inspecting a sapphire. Like the Presidential Commission of Good Government, the challenge the great fictional detective faced was not finding the jewel but proving its connection to criminal activity. Because where there is a great gemstone, wrongdoing is surely close behind: 'Of course it is a nucleus and focus of crime. Every good stone is. They are the devil's pet baits. In the larger and older jewels every facet may stand for a bloody deed.'[xiii]

Why do our consuls and our praetors go about
In scarlet togas fretted with embroidery;
Why are they wearing bracelets rife with amethysts
And rings magnificent with glowing emeralds;
Why are they holding those invaluable staffs
Inlaid so cunningly with silver and with gold?

Because barbarians are coming today;
And the barbarians marvel at such things.

C.P.CAVAFY
'Waiting For The Barbarians'

SACRED
STONES

AMETHYST

CINNABAR

CAIRNGORM

GLOBIGERINA
LIMESTONE

JADEITE

GRANITE

JET

SARSEN

PELE'S HAIR

TUFF

TURQUOISE

SACRED STONES

THE STEADFASTNESS OF STONE SUGGESTS UNIMAGINABLE expanses of time spreading out behind and before us. This quality, of life beyond the mortal span, is one we humans attribute to our presiding spirits, be they gods or demons. Our earliest deities were inscribed in the landscape: mountains, caverns, gullies and cliffs, in the mysterious forms of glacial erratics, meteorites and eroded sandstone archways.

The stone structures and artefacts passed down to us from our earliest ancestors are generally assumed to be sacred. Why go to the effort to haul monoliths across country, chip beads out of blue stones, raise pyramids in the desert, dig subterranean temples, polish jade to dress an esteemed corpse, if not to commune with supernatural forces?

Across history, specific stones have been venerated in their own right. In the Eastern Mediterranean, baetyl were cult stones animated by the spirit of a deity (many are now thought to have been meteorites). After the patriarch Jacob dreamed of a ladder carrying angels between heaven and Earth, he anointed the stone he had used as a pillow, and named the place Beth-el – house of God. The Greeks venerated the navel stone – omphalos – at Delphi, a baetyl marking the centre of the world, the point at which two eagles released by Zeus from opposite ends of the Earth crossed one another's path.

In Jerusalem, the deeply gullied slab of limestone sheltered by the seventh-century Dome of the Rock is sacred within Islam as the site from which the prophet Mohammed ascended to heaven for an encounter with God. Within Judaism, the stone is also identified with the Foundation Stone on which the world was created. Ancient, and perhaps heavenly origins are likewise attributed to the Black Stone of the Ka'bah, the eastern cornerstone of the cube that stands at the centre of the Masjid al-Haram at Mecca.

The earliest known temple was erected at Göbekli Tepe in south-east Anatolia some 11,500 years ago, during the long transition between Neolithic hunter-gatherer societies and the earliest settled agriculture. Göbekli Tepe – 'navel hill' – is believed to have been a site of pilgrimage and ritual, including funerary rites, and perhaps animal sacrifice. All these thousands of years later we still erect stone monuments to glorify our gods, and raise lithic markers over our dead. In Tibetan Buddhism, individual *mani* stones inscribed or painted with the mantra *Om mani padme hum* are placed one alongside another to form mounds or walls. Pebbles – marking the endurance of memory – are left after a visit to a Jewish grave.

What is magic but the unsubstantiated miracles of another's faith? Long condemned, nature-centric faiths – including witchcraft and paganism – are now among the fastest growing religions in the West. In the late 1960s, nature writer and maverick environmentalist Edward Abbey suggested that our increasingly pagan society was 'learning finally that the forests and mountains and desert canyons are holier than our churches' and suggested that we venerate the natural landscape as we might any place of worship. As to his own faith? As Abbey put it: 'I'm not an atheist: I'm an earthiest.'[i]

#01

AMETHYST

THE WALLS OF THE CHURCH OF SÃO GABRIEL IN AMETISTA DO SUL, Brazil, coruscate with crystals, some pale as lavender others brooding violet. Curved, bisected geodes form sparkling arches over saints and relics in niches around the chapel. One vast crystal-lined rock glistens dark indigo beneath the altar, and two others, slim and columnar, support candles to either side. Behind them the painted figure of Christ walks through a dusty valley. It may illustrate a gospel story, but the vale in the painting looks like the area just outside town where a channel has been dug down to the basalt bedrock exposing routes into the crystal-rich rock beyond. Many of the fragments that make up the forty tons of amethyst lining this new church have been donated by parishioners: most, indeed, have probably been mined by them. The design of São Gabriel celebrates the fruits of the earth, and in Ametista, the self-proclaimed amethyst capital of the world, those fruits are crystalline and plum dark.

Ametista do Sul became a city of that name in 1992: previously the parish was known, like the church, as São Gabriel. Colonised from land inhabited by the Kaingang people at the turn of the twentieth century, São Gabriel was a small farming community. In the 1930s and 1940s locals started to find semi-precious stones embedded between the roots of trees and in riverbeds. As fascination with spiritual and occult practices took hold in Europe and the Americas in the 1970s, so, too, did the global trade in semi-precious stones. There had always been a market for amethysts – as geodes for mineral enthusiasts and gemstones for jewellery – but thanks to New Age mysticism, crystals acquired a lucrative new market.

As ever, it was less lucrative for those mining crystals than for those glorifying and trading in them. Amethyst is quartz – most common of all rock-forming minerals – turned purple in the subterranean depths by a dash of ferric iron and a soupçon of radiation. Large, impeccable geodes and rare formations are worth thousands, but you can pick up a rough amethyst for pocket change. The real profit is in processing and performance: turning these raw crystals into products or elaborating their mythos. Alongside the New Age shops, healers and purveyors of general feel-goodery in the 1970s and 1980s, a literary genre arose guiding converts through the many health benefits and wondrous applications of crystals. According to the mineralogist and feminist author Barbara G. Walker, who routinely used crystals to focus her mind for meditation, this was a genre to be filed strictly under fiction. She was riled by the cynical profiteering and dangerous quackery of this new industry. Most of Walker's *Book of Sacred Stones: Fact and Fallacy in the Crystal World* (1989) is dedicated to debunking crystal mysticism. Amethyst, she notes archly 'is a busy healer', purportedly able to 'treat' eye problems, hypoglycaemia, dyslexia, headaches, inebriety, alcoholism, gout, diabetes and urinary trouble, able to cure any skin disorder, dissolve blood clots and absorb negative forces.[i] She is magnificently scornful, busting pseudoscience, pointing out errors of geology, physics and even simple numeracy. She also traces many of these New Age beliefs back to the pages of medieval lapidaries and further. Informing us that one modern crystal mystic claims amethyst 'transmits the violet ray from the planet Mercury' you can almost hear her sigh in exasperation: 'Indeed we have not come far from the old days.'

The specific belief that amethyst was a talisman against intoxication certainly has a long history, and a more tangled one than Walker allows. The stone's name is popularly believed to derive from the Greek *a-methustos* – 'not drunken'. Not all accept the etymology, the first-century historian and philosopher Plutarch among them, who argued that amethyst was merely named for its vinous colour.[ii]

Four hundred years later, the link between amethyst and inebriation reappears in Nonnus' *Dionysiaca* (c. 450 CE) an epic account of the early life of Dionysus, god of the grape harvest, wine, fruit and fertility. Beards soaked in wine, the satyrs partying with the young god become messy drunks and start making unwelcome lunges at the nymphs. Nonnus attributes Dionysus' relative composure to an amulet presented by the great mother goddess: 'To Dionysos alone had Rheia given the amethyst, which preserves the wine drinker from the tyranny of madness.'[iii] Rheia's amethyst bestows on wine-loving Dionysus the vaunted quality of being able to 'hold' his drink.

Another story links Dionysus – in his Roman identity as Bacchus – to the origin of the stone itself. Here the intoxicated god pursues a nymph named Amethyste who prays to Diana to preserve her chastity. The goddess transforms her into a white stone, over which the remorseful Bacchus pours his wine, dying her purple and bestowing on her the power of sobriety. The story has overtones of Ovid's *Metamorphoses* but actually dates to sixteenth-century France. Rémy Belleau's 'L'Amethyste, ou Les Amours de Bacchus et d'Amethyste' features a distinctly Christian sounding Bacchus in the guise of young lover, regretting his behaviour as a drunken fool. As recompense he gifts the world a talisman to escape the 'disturbance of the brain by vinous passions':

> *Now fed by my divine anger*
> *And tinted with my colours*
> *I wish for this fine stone*
> *To guard its carriers*
> *Against the intoxication of my sweet liquors.*[iv]

Walker makes a spirited case for amethyst's sobering qualities being a grand scam on the part of mean hosts and wily servants. She imagines wine being served in goblets carved from amethyst, and the purple stone disguising watered-down wine. She also suggests that amethysts became known as the Bishop's Stone in medieval Europe for a similar reason: senior churchmen could maintain a level head disguising their abstemiousness behind a carved purple cup. It's a charming idea, but there are no amethyst vessels of appropriate vintage in the vast collections of either the British Museum or The Metropolitan Museum of Art.

In the *new* New Age – spirituality, magic and paganism translated into personal wellness for the twenty-first century – demand for crystals is again high, and serious money is being spent. In 2017, a specialist dealer described crystals as a 'billion-dollar industry, easily'.[v]

Amongst the other properties now assigned to it – mitigating migraines, soothing sleeplessness, neutralising negativity – modern mystics hail amethyst's power to combat overindulgence and help with weight loss and mindfulness. It's a far cry from Dionysus being endowed with the power to keep drinking and never leave the party.

Metres of newsprint have been dedicated to debunking mystic claims attached to luxury crystal products – amethyst-infused water bottles, yoni eggs – promoted by celebrities such as Gwyneth Paltrow. Close behind have come investigations into the murky side of the crystal business, which includes perilous mining conditions and child labour. Reporting for the *Guardian*, Tess McClure visited the town of Anjoma Ramartina in Madagascar, known for its crystal exports to the US, and sat down with the local major Many Jean Rahandrinimaro. 'He placed a few stones on the wooden table in front of him: polished clear quartz and purple amethyst. He estimated that from from a population of about 10,000 people, up to a quarter of locals now depended on the mines for some income'. Between two and four men die in the crystal pits surrounding the village every year.[vi] Dangerous, unregulated and often polluting, crystal-mining practices hardly chime well with the good-vibes-only, Earth-mama image of wellness culture.

Back in Ametista do Sul the economy has started to diversify away from dependence on the crystal mines. Blessed with volcanic soil and well-excavated caves with a stable temperature of 17 °C, local entrepreneurs have engaged in a new industry: they're making wine.[vii]

02

CAIRNGORM

THERE ARE PECULIARITIES HIDING IN THE DIM, WOOD-LINED Enlightenment gallery at the British Museum. This long chamber is dedicated to the early men of science, their conception of a world governed by rules and logic, a measuring stick for the distance they could place between themselves and 'primitive' superstition. It's a space evoking a culture so patrician, so collegiate, one anticipates the smell of brandy and bridle leather. Yet, even here, magic has snuck in.

One case holds a mirror of black obsidian, pale wax discs inscribed with arcane symbols, a gold circle, and a polished sphere of smoky quartz – a cairngorm, known by the name of the Scottish mountain range in which it's found. All were once the property of sixteenth-century mathematician, astrologer, philosopher and alchemist John Dee, a scientist who considered his greatest work to be the product of conversation with angels.

Dr Dee recorded the particularities of his own birth with an astrologer's precision: 4.02 p.m. on 13 July 1527, fifty-one degrees and thirty-two seconds north of the equator.[i] He came of age during the time of King Henry VIII, nearly didn't survive the reign of Queen Mary, but entered that of Elizabeth I on a good footing. Elizabeth prevailed on him to select an auspicious time and date for her coronation. His calculations suggested 15 January 1559. As a Protestant and a woman, much rested on the spectacle of Elizabeth's coronation: the trust placed in Dee's astronomical expertise is a mark of the esteem afforded him and his technique.[ii] During this period of favour, Dee was consulted on many subjects, from the appearance of a new star to the maintenance of the navy.[iii]

These were febrile times, and dangerous for scholars exploring what were then obscure branches of knowledge. The new Protestant establishment condemned Catholicism – with its miracle of transubstantiation and sales of indulgences – as magical. Accusations of either popery or conjuration carried mortal consequences. Even Dee's mastery of theatrical mechanics – a beetle, made to 'fly' across the stage – gave rise to whispers of occult powers.

There was barely a line separating acceptable and unacceptable practices. European monarchs were fascinated by alchemy and the gold it might bring, but that, already, was sailing close to the wind.

Dee regarded the consultation of crystal balls – showstones – an extension of his scholarship in other fields. Just as he studied the heavens through the lenses of a telescope, so this other glass might make visible that not evident to the naked eye on Earth.[iv] Crystals were thought to be pure, so they were able to attract the purity of spirits and indeed entrap them. Dee considered what others termed magic to be the human ability to tap into the divine force governing the movement and behaviour of all things under heaven.

A desire to understand this great unifying principle prompted him to attempt to communicate with angels.[v] Receiving visions through a crystal – scrying – was not a skill available to Dee. Instead his angelic conversations were conducted through a succession of young men – often peculiar, marginal figures – paid for their services.

The most significant was Edward Kelley, who first came into Dee's services under the name of Talbot. He arrived at Dee's house at Mortlake in March 1582 'to see or shew some thing in spiritual practice'. Dee was, with good reason, suspicious of 'Talbot' and made it plain that he did not study or exercise 'that vulgarly accounted magick' but was instead desirous of help in his 'philosophical studies, through the company and information of the blessed angels of God'.[vi] During their first 'action' together, Kelley – or Talbot – summoned a vision of the good angel Uriel.[vii]

Kelley was then twenty-six, nearly thirty years Dee's junior. A mercurial figure prone to violent bursts of temper, he walked with a stick and wore his hair long to hide a cropped ear – punishment, perhaps, for some misdemeanour. He had been accused of forgery and coining, among other things. Dee forgave extraordinary transgressions by Kelley, including reading and tampering with his private diary. In return the younger man provided him with private spectacle, a series of angelic messages delivered over seven years, the substance of which were recorded by Dee in many hundreds of manuscript pages.

Kelley's revelations to Dee were theatrical, delivered in apocalyptic language, performances for an audience of one. At their apex sat the crystal ball. The visions commenced with the sight of a black curtain being opened, and concluded as it fell. Dee possessed at least three different stones: a '"great Chrystaline Globe", a crystal sphere mounted in a wooden frame, and a stone that the angels deliver to him.'[viii] This last was the smoky, peat-tinted cairngorm, which came to Dee in a piece of drama orchestrated by Kelley in their first action together after a break of six months. Kelley had summoned a spirit called King Camara, and Dee confessed to the spirit his struggle interpreting information conveyed to him. 'One thing is yet wanting,' the King told him. 'A meet receptacle . . . a Stone . . . One there is, most excellent, hid in the secret of the depth & c., in the uttermost part of the Roman Possession. Lo, the mighty hand of God is upon thee, Thou shalt have it. Thou shalt have it. Thou shalt have it. Dost thou see? Look and stir not from thy place.'

Dee was told he would prevail with this stone 'whose beauty (in virtue) shall be more worth than the kingdoms of earth. Look, if thou seest: but stir not, for the angel of his power is present.' Then, by the study window Kelley beheld an angel the height of a young child, who held out a stone to Dee 'as big as an egg, most bright, clear and glorious'. Through Kelley, Camara instructed Dee to go toward the angel. He could see nothing, but noticed a shadow on the ground 'roundish and less than the palm of my hand. I put my hand down upon it, and I felt a thing cold and hard' – the stone. Camara instructed him to 'Keep it sincere . . . Let no mortal hand touch it, but thine own. Praise God.'[ix]

This was the showstone with which Dee and Kelley were later to travel across Europe, variously feted and hounded. It was also the stone from which Kelley, fatefully, received a vision at Trebon Castle in Bohemia instructing the two men to 'hold their wives in common'.[x] Despite Dee's reluctance he consented to the 'doctrine of cross-matching' and somehow persuaded his wife Jane to participate in an angel-appointed wife swap. The son born to Jane nine months later was called Theodorus Trebonianus – the gift of God at Trebon.[xi]

Such visions seemed to manipulate Dee to Kelley's advantage. It is impossible to tell at nearly five centuries distance how sincere the scryer was in his enterprise, and to what extent the wise man was played for a fool. Dee was aware of Kelley's flaws as a man, and from the early days of 'Talbot', suspicious of subterfuge, yet he had great faith in the substance of the visions themselves – the messages emanating from the pure stone blessed by God.

For centuries, Dee's angelic conversations condemned him to the margins of history. More recently, his scholarship has been appraised in the context of a time when magic and mathematics were considered sibling arts. In his book, *The History of Magic*, Chris Gosden draws parallels with a scientist born a generation after Dee – Isaac Newton – who searched for a unifying theory of matter, studied waves from the heavens and the refractive qualities of crystal, and who also dabbled in alchemy. John Maynard Keynes had argued that: 'Newton was not the first of the age of reason. He was the last of the magicians, the last of the Babylonians and Sumerians.'[xii] Perhaps Dee's cairngorm is in its rightful place among these trappings of the Enlightenment, and superstitions not so far removed.

#03

CINNABAR

C INNABAR IS NOT ONLY A BEAUTIFUL RED — AS A COMPOUND OF
sulphur and mercury it is the closest manifestation in nature
of that mythic substance, the philosopher's stone (see also: Sulphur,
p. 295). Since the third century BCE, Chinese alchemists sought
the elixir of eternal life. Li Shao Chun, an alchemist in the court
of Qin Shi Huangdi, first emperor of unified China, described
how cinnabar might be transmuted into gold: made into utensils
for eating and drinking, it would prolong life and prepare the route
to immortality.[i]

In eighth-century Iran, the great alchemist of the Islamic Golden
Age, Abu Musa Jabir Ibn Hayyan (also known as Geber in Europe),
introduced the principle that all metals were formed in the earth
from sulphur and mercury in different proportions. For centuries
afterwards, alchemists considered the principle ingredients of
the 'philosophers stone' to be alchemical mercury and alchemical
sulphur.[ii] Mercury was sublimated from raw mineral cinnabar, then
combined with sublimated sulphur to create a pure alchemical
form – known as vermilion in Europe, and *hingaloo* in Sanskrit –
considered to have extraordinary spiritual, medicinal and transformational potency.

The Indian practice of alchemy – *Rasayana* – developed between
the seventh and fourteenth centuries CE, and was influenced by the
spread of ideas from China and the Arab world. Sanskrit treatises
on Ayurvedic medicine describe various compounds containing
sublimated cinnabar destined for ingestion by the patient. As well
as elixirs for the flesh and spirit, cinnabar was used as a devotional
mark on the forehead, and married women indicated their status by
applying it to the partings in their hair. Artists, too, used cinnabar
as a pigment making paintings that were powerful and sacred even
in their materials.[iii]

It can't have taken many accidental poisonings for the toxicity of
mercury to manifest itself: both physician and artist needed to learn
how to mitigate this effect while still using cinnabar as a remedy or
pigment. Instruction manuals on painting from India and Iran in the

sixteenth and seventeenth centuries advise artists to add milk, lime juice and honey to cinnabar as they process it. Some ingredients may have enhanced the texture of the paint, while others were considered antidotes to its poison.[iv]

Through *Rasayana*, the alchemist sought the transmutation of body and soul to higher planes: as such, it was closely related to the esoteric practice of Tantrism.[v] In alchemical *Tantras* (instructional texts) of the tenth century, the idea emerged that alchemical mercury was divine semen – the emission of Shiva, burning hot as it projected from his body – and alchemical sulphur was the fragrant menstrual blood of the mother goddess Devi.[vi] Cinnabar thus symbolises their sexual union.

The offspring of this alchemical tradition endures in *Pichhvai* painting, a practice that requires a lifetime of devotion to master. The journey starts with an arduous year-long apprenticeship in which the young artist learns to make paint from stone. *Pichhvai* means 'hanging at the back': these large paintings are exquisite backcloths suspended behind the idol of Shrinathji – the god Krishna manifested as a seven-year-old boy. Each painting drama-tises an episode from the god's life – when he lifted the Govardhana hill to save the city of Vrindavana from a devastating flood, perhaps, or when he replicated himself so that he could partner every milkmaid (*gopi*) in a dance. The paintings are changed according to a cycle of seasons and festivals, then rolled up and stored.

Pichhvai has its origins in seventeenth-century Rajasthan,[vii] and shares its techniques with Persian miniature painting and its symbol-laden geometry with Buddhist *thangka*. Slow and precise, it is a sacred practice in which the artist must yield individual self to collective enterprise. Every aspect, from the proportions of the figures to the materials used in the paint, has symbolic signifi-cance.[viii]

In the *Pichhvai* studio raw cinnabar is ground in a granite mortar shaped like a vulva, and the action of the phallic pounding tool becomes a generative force. Seated cross-legged on the floor, an apprentice aligns the mortar full of minerals with his base chakra as he works. The process can take two weeks, with the pigment repeatedly crushed, washed and filtered. It is laborious: many apprentices don't last the year. The result is a fine powder, stored in folded paper envelopes like jewellery. For paint, it is mixed with a little distilled water and gum Arabic. A base layer of intense background colours is painted first, then burnished to a shine with agate, presenting a silken, grit-free surface to the brush for the fine work that will follow: powdered stone layer upon powdered stone layer.

Each pigment being derived from raw mineral, the final painting in effect becomes a sheet of refined, silken stone: malachite for pale green, lapis lazuli for ultramarine, red and yellow ochre for earth tones, and intense scarlet from cinnabar. The apprentice working the pigments – transforming raw stone from a base to a subtle material – is re-enacting a practice derived from *Rasayana*.[ix] The use of mineral colours honours the Earth as a living soul, with the artist now assuming the role of the alchemist.

04

GLOBIGERINA LIMESTONE

I N THE ENTRANCE OF THE TARXIEN TEMPLE IN MALTA STANDS THE
lower half of a giant figure, carved deep into the rock. Rising from
delicate ankles, its calves billow out like fleshy melons. Beneath
pleated skirts, the lush roundness of the sculpture's thighs and
hips blossom to the dimensions of a small (if generously padded)
armchair. Carved around 3100 BCE, at the height of a flourishing
temple culture on the islands that lasted over a millennium, our
fleshy friend would once have stood almost three metres high: an
imposing emblem of plenty.

Unearthed in the early 1900s, this sculpture at Tarxien is the
largest surviving example of prehistoric cult objects from Malta.
Dubbed the 'fat ladies' they include small totems left as grave
gifts, and sculptures of seated and sleeping figures, as well as the
monumental bodies carved into the stone of the temples themselves.

The twenty-eight Neolithic temple complexes on Malta are the
world's oldest surviving freestanding structures.[i] The earliest of
them pre-date Stonehenge and the Great Pyramid at Giza. Malta's
globigerina limestone is golden and fine-grained – an accretion of
shells from tiny *globigerina foraminifera* plankton that once sank
to the seabed off what is now the north coast of Africa. The
softness of the stone helped ritual art and architecture flourish on
the islands between 3300 and 2450 BCE. Then, mysteriously, after
over a thousand years of increasingly sophisticated artistry and
ritual, the temple-building culture died out in around 2000 BCE.

In the world of stony things, globigerina limestone is easy to
quarry and carve, though forty-five-ton blocks required formidable
skill. Visitors in the seventeenth century were told the temple ruins
were evidence of a race of giants who lived in Malta before the great
flood. One temple complex on Gozo carries the old legend of the
stonemason giants in its name: Ggantija.[ii]

Within the women's liberation movement the possibility that
one of the oldest sophisticated cultures in the Mediterranean
venerated full-bodied female figures was irresistible. In the early
1970s, feminist theory suggested that Neolithic people were gentle,

matriarchal goddess-worshippers living in harmony with the earth: the violent patriarchy that replaced them came with the metal tools and weapons that followed.

A leading force in this goddess movement, archaeologist Marija Gimbutas described the stone temples and sculptures of Malta in terms of feminine fertility.[iii] Gimbutas identified egg shapes everywhere, from the robust hips of the statues, to decorations on the pedestals they stood on and the shape of the temple chambers. The characteristic curving forms of Maltese temple architecture suggested womb-like spaces, as did the curved bodies of the 'fat ladies': all entirely appropriate for the rituals of death and regeneration that would have taken place there.

Malta is a focal site for the goddess movement, but alas the credibility of a single, dominant matriarchal goddess cult depends on Malta's 'fat ladies' being female. This is not always the case. Caroline Malone, one of the lead archaeologists on the temple complexes, points out that while some statues do have large breasts and female genitalia, and one small figure appears to be pregnant, most 'fat ladies' have no evident gender markings of any kind. Belief that they were female was based on the sculptures' rounded hips and thighs, and their pleated skirts. Malone suggests that rather than being fat ladies, the sculptures are un-gendered: symbolic ancestor figures shown in a state of idealised abundance.[iv]

That interest in abundance offers a clue as to why the temple culture disappeared. Malta today is dry and rocky, but it was not always so. When the ancestors of these prehistoric stone carvers travelled from Sicily with their sheep, goats and oxen over 6,000 years ago, the landscape that greeted them was verdant. The islands were soon deforested. As the population grew over the centuries, and agriculture became more intense, the soil was eroded and the land became less fertile.

Studies of bones and geological samples show that the mania for sculpting corpulent figures and building temples was at its strongest as living standards declined. Rather than looking to the world beyond Malta, or developing the agricultural infrastructure within it, the focus turned ever-more to honouring insular cults. It is, as the archaeologists suggest, 'a cautionary tale about what happens when a people focus too much energy on worshipping life rather than sustaining it.'[v]

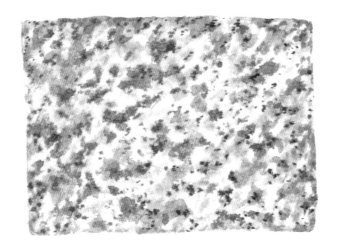

#05

GRANITE

G REAT GRANITE MONOLITHS STAND OUT LIKE GIANT WALKERS crossing the broad, scrubby grassland of the Eurasian Steppe. Many are carved in the stylised likeness of humans, positioned to face the rising sun at the winter solstice. They are cloaked in herds of deer that swirl upwards around them as if in flight, legs tucked and curling antlers lowered along their backs, caught in such concerted movement that they might carry the figure upwards with them. Engraved hoops define the figures' ears, one hoop on the north side, one on the south. A string of beads circles their throat, and around their waists are belts strung with diverse tools and weapons: swords, daggers, knives, fire-starters, chariot rein hooks and quivers.[i] Standing as tall as four metres, few have a recognisable face, but these slab-like standing stones may have represented specific individuals, identifiable to their followers by their weapons and heavily tattooed bodies.

Over a thousand deer stones like these dot the Eurasian Steppe in clustered groups. Most are in Mongolia, with the finest examples in the north-central area of the country. This, significantly, is the most productive grazing land in Central Asia and still home to a tradition of nomadic herdsmen who work the Steppe on horse-back.[ii] Carved and erected during the late Bronze Age – between 1200 and 700 BCE – the stones are the fruit of huge collective enterprise. Twenty people or more would have been needed to transport each of these granite slabs into position, to peck or grind complex and elegant imagery into the surface of the rock with metal tools, and then to dig into the unyielding ground and lever them upright.

Deer stones form the most visible and dramatic elements of large funerary complexes on the open Steppe, located near burial sites that were active for many generations. While they are not grave markers in a literal sense – they are set apart from the actual tombs, and no human remains have been found buried beneath or beside deer stones – these granite menhirs present the human figure caught in a dynamic relationship between two animals – the wild deer and the domesticated horse – that guide its passage through and beyond life.

The tribes of the Eurasian Steppe were the earliest horse people. The first signs of domestication of the animal date back 5,500 years, to the Botai culture in what is now northern Kazakhstan. Horses were well adapted to the climate and geography of the Steppe: they thrive on dry grassland, need less water than cattle, forage over great distances, and can survive harsh winters. Their hooves allow them to dig for water in a dry summer, and to break through crusted snow to access vegetation in the coldest months.[iii] Horses were bred for meat, milk and hides. At some point they were used to pull wheeled vehicles, and at some point they were trained to accept a human rider seated on their back (there is an unresolved cart-before-the-horse debate as to which occurred first).[iv] Thanks to the deer stones, and the structures surrounding them, we do know that by 1200 BCE, there was already a sophisticated nomadic riding culture on the Steppe.

Horse transport was transformational: it allowed humans to cover great territory at speed, spreading their genes, disease and culture as they went.[v] People developed a new relationship to landscape in this expanded and fast-moving world: the huge grassy plains needed to be navigated, divided, delineated. These horse-riding nomads began to create new kinds of monuments, structures that could be seen at a distance, and which became gathering places, sites of ritual connected to transformative episodes: life, death and what lay between.[vi]

Like the Scythian cultures that succeeded them, these early nomads were often buried with their horses. Horse sacrifice was an important part of their ritual culture: at about thirty sites, deer stones have been found encircled by burial mounds of horse remains, usually bundles of heads, hooves and cervical vertebrae. They have been interred with great ceremony, with each skull positioned toward the east: like the deer stone, they are facing the rising sun.[vii] Recent analysis of these remains has found wear to the teeth and cheekbones: these horses had been bridled with a hard bit, and heavily exerted.[viii]

And what of the deer? While other wild animals appear in early nomadic art, the deer dominates. The Iron Age Saka-Scythian people who lived in what is now eastern Kazakhstan wore fine gold plaques in the form of both male and female Altai maral deer, sewn all over their outer garments. Horses found in their burial complexes were dressed in magnificent costumes complete with wooden horns, transforming them into deer for their final trip with their riders.[ix] Art historian Esther Jacobson-Tepfer has identified a commonly held mythic tradition in which a female deer-like figure appears at the centre of cosmic processes: a deer goddess, who played a central role in the spiritual lives of peoples in south Siberia and Mongolia over several millennia.[x]

Other studies have noted the beak-like mouths of the deer circling the stones and suggested they embody a moment of spiritual transformation: a shamanic flight from the Earth to the heavens (see also: Red Ochre, p. 210). If the stones were intended to embody actual people, and the engravings their tattoos, then it seems these early nomads covered their skin in protective images of deer, or maybe hybrid deer–birds, or emblems of a deer goddess.

These hard-worked granite monuments were not passive representational objects. They had their own power, derived in part from the resistant stone that formed the granular bedrock of the landscape they stood in and which spoke to the significance of that specific location. The deer stones may have played multiple roles for the early nomads, and have meant a variety of things: possibly their significance shifted over the centuries. Vertical markers in a wide horizontal landscape, they embody transformation from one state to another. They stand surrounded by horses, greeting the living at the site of the dead, and the dawn after the darkest day of the year.

#06

JADEITE

THE MAIZE GOD WAS THE IDEAL OF MAYA BEAUTY: VIGOROUS, young and green. His thick hair flopped over his face in hanks like corn silk. He was tall and elegant like a corn stalk. Often he is depicted as a contortionist, young and limber, his body sprouting above him like a world tree. Every harvest, the Maize God was decapitated and consumed by the Maya people, regenerating the following season. In the *Popol Vuh* – the sacred book of the indigenous Quiché Maya people of Central America – humans are born from maize.[i] It is the stuff of life – point of origin and staple diet – and the maize plant furnished the Maya with a concept of divine balance and cycle of resurrection.

To say then that jade was venerated by the Maya for its association with ripening maize is no faint praise: in a world without metal tools, it was prized above all other materials, furnishing axe-heads, jewellery, devotional figures and funeral masks. In the gemological world, 'jade' is used for two different stones, both of them green, metamorphic, fine-grained and among the toughest substances on (or in) Earth. Nephrite is ancient Chinese jade, and jadeite the green stone of the Maya. It would have been hard-won in the Maya world. The only source at that time was the Motagua River valley of Guatemala, and while it could be found in colours from purple to green to hazy white, it was the bright green and deep blue varieties the Maya most prized. Fracturing, sawing, drilling, grinding and polishing this unyielding stone would have been arduous.[ii]

The Maya have long roots, but at the height of their power – 250 to 900 CE, known as the Classic Period – they occupied the territory of southern Mexico, Guatemala, northern Belize and western Honduras. Their reverence for jadeite was ancient, inherited from the Olmec (c. 1200–400 BCE), if not earlier: it's durable stuff, a stony green link to venerated ancestors. From the Olmec, the Maya took the practice of placing a jade bead in the mouth before burial, symbolising the maize kernel that would bring renewal. Maya understood their connection to the Earth to be reciprocal. You could not simply extract building materials, precious

stones, minerals and food: you also had to give back, with offerings
up to and including human sacrifice.

All things also had an animating spirit that needed to be tended.
The construction of a new house, for example, required a
dedication. First, offerings were buried in its foundation, perhaps
the heads of sacrificial chickens, precious ceramics and jade
carvings. A shaman would then perform a rite to compensate the
Earth Lord for the materials he had provided, and summon the
ancestral deities to provide the house with a soul. At the end of the
building's useful life, a termination ritual would be performed in
which ceramic and jade objects might be broken, the house partially
destroyed, and incense burned, to release its soul.[iii]

Dense and cool, quick to gather condensation and release
vapour, jade had a close association with the breath spirit that
animated the living human body.[iv] Carrying forward that association with coolness and moisture, jade jewellery – particularly
'flares' worn on or passing through the ears – was often carved in the
shape of moving vapours representing breath, the winds, and the
perfumes and music that might be carried on them.

For the Maya, jade embodied the quality *yax,* which described
the blue-green zone on the colour spectrum, and symbolised water,
new growth and ancestral connections. *Yax* was also embodied by
the intensely pigmented paint blend known as 'Mayan blue', and
the iridescent plumes of the quetzal bird, worn by kings and paid as
tribute. In Maya cosmology, the sacred 'tree of life' or 'first tree'
(a mighty ceiba) was known as *yaxte.*[v]

The Maize God displayed *yax* in abundance. As an eternal
youth, he offered an androgynous ideal of beauty, elegance and
plenty, and his costume inspired dress for men and women. He
drips with jade: a diadem and headband (with quetzal feathers),
a necklace and face pendant, a beaded overskirt, jade bands for his
wrists, and jade ear flares. In this finely worked carapace of green
stone he would have rung like a bell as he moved, as would those
who dressed in imitation. Such finery would have been beyond the
means of all but the elite. The Mayan ruler K'inich Janaab Pakal,
king of Palenque[vi] for sixty-eight years (615–683 CE), is depicted
forever young and beautiful, dressed as the living embodiment of
the Maize God. He was buried with a green jade funeral mask,
carrying the association of eternal regeneration into the grave.

The Maya integrated the *yax* of jade into their own bodies through piercings. The last word in Maize-God chic was jade inlays in the front six teeth. The process would have been long and painful, with the dentist grinding the tooth surface with a thin tube (jade in the early years, later copper) powered by hand or a bow drill. A paste of powdered quartz was used as an abrasive beneath the drill to cut a perfect round hole in the enamel and dentine.[vii]

As the *Popol Vuh* warned, such finery carries the danger of pride. The false god Seven Macaw adorns himself in riches and proclaims himself to be the sun and moon: 'My eyes sparkle with glittering blue/ green jewels. My teeth as well are jade stones, as brilliant as the face of the sky.'[viii] He is shot in the jaw with an arrow from a blowgun by the Hero Twins, and crippled with pain. Duplicitous healers, sent by the Hero Twins, attend to Seven Macaw, and tell him his teeth must be removed. He resists, despite the pain: 'It is perhaps not a good thing that my teeth come out, for it is only because of them that I am lord. My teeth, along with my eyes, are my finery.' Nevertheless, the healers pluck out his teeth and eyes, and he dies, resembling a god no more: 'Thus the wealth of Seven Macaw was lost, for the healers took it away – the jewels, the precious stones, and all that which had made him proud here upon the face of the earth.' You may dress like a deity, green teeth and all, but all the jade beauty in the world cannot make you the Maize God.

#07

JET

I N APRIL 204 BCE, DURING THE BRUISING SECOND PUNIC WAR, a small, black, pointed stone was carried triumphantly into Rome. This was not a spoil of war: in accordance with a prophecy, the stone had been fetched from Asia Minor by a delegation selected from five important families of Rome. The worthiest citizens were appointed to meet it on arrival. Women passed the stone safely from boat to shore: an object so potent that the matron Claudia Quinta whose reputation, according to Livy, was 'previously rather shaky', hereafter became celebrated as an icon of virtue for her role in its secure delivery.[i] The black stone was temporarily placed in the Temple of Victory, and its arrival celebrated with the inaugural Megalesian games.

The war was fourteen years in, and Rome had endured humiliations and catastrophic losses. Things had started badly in 218 BCE, when Hannibal of Carthage had led his army – complete with war elephants – up the Iberian Peninsula, across Gaul and crossed the Alps to launch a surprise attack. In 205 BCE the beleaguered Romans called for divine support. The Sibylline Books – prophecies only turned to at moments of great crisis – were consulted.

According to the Sibylline prophecy, to expel the enemy from Italy, the Idaean Mother should be fetched and brought back to Rome.[ii] This was the black stone: a *baetyl* (or 'bet el' – 'house of god') imbued with the spirit of the Anatolian mother goddess Cybele, known to the Romans as *Magna Mater*, the Great Mother – also the Idaean Mother since Mount Ida near Troy was sacred to Cybele. This association with Troy played into a popular foundation myth of Rome: Aeneas and other heroes of Troy escaped the destruction of their city and sailed to Italy, where they furnished thirteen generations of fictional kings (to fill in a few inconvenient centuries) before the last of the royal line, twins Romulus and Remus, founded the city on the Palatine Hill. Far-fetched as the tale was, it filled the Romans with fighting pride to imagine themselves descended from Trojan warriors, and fighting pride was what they needed to defeat Hannibal.

Cybele worked her magic, Hannibal was expelled, and the Asian goddess was honoured as defender of the city. The cult of Cybele was widespread in Anatolia during the first millennium BCE but its introduction into Rome was not uncontroversial. Cybele was attended by the *galli* – priests who castrated themselves in memory of the goddess's lover Attis who had cut off his genitals and bled to death. Depending on the story, he was either punishing himself for cheating on Cybele with a nymph, or ruining himself to avoid forced marriage.

Engaging in ecstatic rites including self-flagellation and slashing themselves with knives, the *galli* adopted feminine dress and curled their hair, wearing perfume, jewellery and headdresses. 'The Romans felt a profound unease towards these self-castrated men who by their very existence challenged the Roman ideas of superiority and social privileges based on masculine gender.'[iii] The acceptable aspects of the cult were described as 'Roman', while those found distasteful were dismissed as 'Phrygian'.[iv] For centuries, Romans were forbidden from castrating themselves and becoming *galli*.

In 1982, during excavations at Cataractonium – a third-century Roman military base on the River Swale in North Yorkshire – archaeologists discovered the tomb of a lavishly bejewelled individual. Skeleton 952 was dressed with an elaborate multi-strand necklace of jet beads, a bracelet of carved black shale on one arm, and of more jet beads on the other. It wore a twisted metal cuff around its ankle, and had been buried with two pebbles placed in its mouth. 'Naturally the assumption was that this was a female,' recalled Hilary Cool, an archaeologist working on the finds. When the osteological report came back identifying the individual as a male aged twenty to twenty-five years, the reaction was that there must be some mistake. The bones were sent out for a second opinion. This concurred with the first. So, in as far as it is possible to be sure, it seems that this was, indeed, a young man.[v]

Jet jewellery was popular in Roman Britain, where the stone was mined and gathered on the north-east coast, near what is today the town of Whitby. Planks of it occur within the stone of the cliffs, and pieces can be found washed up on the beach. Jet is the fossilised remains of driftwood from the Lower Jurassic, deposited 183 million years ago. During this period dramatic climate change caused bacterial and algal blooms in the ocean, leading to toxic conditions and mass extinction (see also: Black Shale, p. 26). These same conditions were responsible for the 'jetonisation' of the driftwood, resulting in lustrous, smooth black material that is remarkably stable, if potentially flammable.[vi]

According to Pliny the Elder, Jet derives its name from a district and a river in Lycia known as Gages. In appearance, it is 'black, smooth, porous, light, not very different from wood, and brittle . . . The kindling of jet drives off snakes and relieves suffocation of the uterus. Its fumes detect attempts to simulate a disabling illness or a state of virginity. Moreover, when thoroughly boiled with wine it cures toothache and, if combined with wax, scrofulous tumours.'[vii]

The magic qualities of jet had, much earlier, been associated by the Greeks with women: jet amulets were recommended to facilitate birth. Protective carvings of gorgons' heads – *gorgoneia* – made from British jet are found throughout the north-west territories of the Roman Empire.[viii]

Skeleton 952's jet beads came from the same Whitby deposits that fifteen centuries later were to furnish Queen Victoria and her contemporaries with black mourning jewellery. Hilary Cool was puzzled: Roman men seldom wore jewellery, and even Roman women did not wear ankle bracelets – who was this mysterious young man, bedecked in black stone finery? Her 'eureka' moment came when she remembered an episode in Apuleius' second-century novel, *The Golden Ass*, in which the hero spends time in the service of a group of eunuch priests of the goddess Cybele.

Recent research had linked the popularity of jet and black shale jewellery in Roman Britain to growing interest in the eastern mystery cults of deities such as Cybele – black stones in honour of the Great Mother's black stone, perhaps.[ix] Cool concluded that Skeleton 952 was a *gallus*: a eunuch priest of Cybele, living in the Roman garrison at Cataractonium dressed in feminine garb and lavishly adorned in jet jewellery. Inscriptions and statuary show Cybele was worshipped in the Roman north: the population of Cataractonium was cosmopolitan, of diverse background and beliefs. Finding a richly attired *gallus*, buried in all of his jewellery, may have been surprising for the archaeologists, but his presence would have been less so to soldiers stationed there in the fourth century.

Back in Rome, by then, the cult of Cybele was in its final years: the ecstatic public rites performed by *galli* made them an easy target for Christians attacking pagan practices. Their behaviour was seen as unbefitting to ministers of god: wanton, immodest, bloody and violent. Voluntary self-castration, above all else, was seen as an abomination.[x] The Christian writers sought to replace one great mother goddess with another. We have not come so far in the intervening centuries. When Cool published her report on jet-bedecked skeleton 952 in 2002, the British press publicised her discoveries in sniggering, scandalised tones that would hardly have been out of place in an anti-pagan screed of the fourth century.

#08

PELE'S
HAIR

PELE IS THE GODDESS OF THE VOLCANO, A VOLATILE FORCE OF creation and destruction. The flowing lava body produced by Pele has formed the islands of Hawai'i, a constant process of birthing and regeneration. She is known as Pele-honua-mea – Pele the sacred earth person – and Wahine-o-ka-Lua – Woman of the crater – and Pele 'Ailā'au – forest-eating Pele. [i] She can be fearsome when roused, but she can also charm and seduce.

Pele's hair is alluring stuff, resembling strands of spun sugar. In the sun its glossy lengths sparkle with golden light. It is formed when lava shoots into the air in fountains, or when bubbles of it burst, sending thin trails of hot volcanic material into cooler air. Rapidly vitrified in flight, these golden brown filaments, some as fine as spider silk, are picked up and carried by the wind, forming great drifts against the billowing black fields of basalt, as though the lava had been garnished by a French *pâtissier*. But trifle with it at your peril. It is pretty but hazardous: fine enough to break the skin and fragile enough to splinter within it. Air-born fragments can damage the lungs, or contaminate water sources.

The composition of lava varies, but it is rich to a greater or lesser extent in silica, the stuff of quartz and window glass. Like obsidian, Pele's hair is a natural volcanic glass formed when silica-rich lava cools too fast to form large crystals. Pele's hair can be found with droplets of darker glassy material known as Pele's tears: one was a splash of lava, the other the fine trail that flew through the air behind it.

These are not phenomena confined to Hawaiian volcanoes, but their identification and naming reflect the deep understanding of volcanic behaviour born of centuries-long coexistence with elemental forces. Geologists use Hawaiian terms to distinguish different crusts of flowing basaltic lava. Pāhoehoe is smoothish, elastic looking, twisty, pouring forward. The sound as it moves is a cindery tinkle, like the sound of sharp crystals being born. 'A'ā has a frazzled, fragmented crust like a rough pile of burning coals, sending vivid red rocks tumbling when it meets a slope.

When geologists describe the formation of Hawai'i they point first to the relative ages of the islands. This is sequential, with the oldest parts of the archipelago at its north west: Ni'ihau, and the fertile, rainforest island Kaua'i are about five million years old. Of the youngest islands in the south east, Hawai'i itself is about 400,000 years old, and with four active volcanoes, still growing and transforming. Forming beyond it is Lō'ihi, a submarine volcano, which may break the surface in 200,000 years.

One theory of their genesis is that the islands have formed over a mantle plume – a duct of hot, solid rock that stretches down almost 3,000 kilometres, deep into the mantle of the Earth. This hot spot creates magma above it, which moves through weaknesses in the rock until it breaches the ocean floor. Here it spews out lava and builds up layers of basalt in bursts, as though exhaling life into the new island to come. The Hawaiian archipelago is in the middle of the Pacific plate, which is moving very slowly – about ten centimetres a year – in a north-westerly direction. According to the 'hot spot' theory, over millions of years the Pacific plate has moved north west over the top of the mantle plume, progressively exposing new sections of the ocean crust to the heat source as it goes. 'The paradise of the Hawaiian islands is thus the product of the most direct route to Hades we have on earth,' writes one European scientist, feeling the need to evoke Greek deities for dramatic effect.[ii]

Hawaiians, of course, have their own ancient gods through which the history of the islands, their volcanoes and the world beyond are described. Pele and her family are the animating spirit of the land, the living rock and water, the elemental forces and the flora, fauna and funga that coexist with them. Their legends are passed down through *mele* (chants), *hula* (dances), and the narrative tradition of *mo'olelo*, a word without adequate translation, but which entwines ideas of history, oral tradition and literature, and in which oratory and locale also play an important role.[iii]

Missionary teachers in the nineteenth century interposed Christian faith and doctrine between Hawaiian people and their volcanic homeland. The holy sites of Christendom, on the other side of the world, were offered in place of the many sacred sites – *wahi pana* – of the Hawaiian archipelago.[iv] The Hawaiian language was banned in schools in 1896, cutting generations off from their parents and grandparents' beliefs and traditions. The dramatic stories of Pele and her kin, their voyage to Hawai'i, the creation of the volcanoes and the living world around them, were re-cast as cute fables for children, divorced both from their sacred associations and the land they related to.

'If Pele is not real to you,' write Hawaiian scholar Mary Kawena Pukui and anthropologist Edward Smith Craighill Handy, 'you cannot comprehend the quality of relationship that exists between persons related to and through Pele, and of these persons to the land and phenomena, not "created by" but *which are,* Pele and her clan.'[v] The oral tradition relating to Pele is vast and includes epic poems. Versions proliferate like the strands of Pele's hair, almost as many as there are tellers. One of the best-loved histories is that of Pele and her youngest sister Hi'iaka.

Pele, daughter of the war god, leaves her homeland Polapola with her kin, riding in the canoe Honuaiākea (earth explorer.) Arriving in the Hawaiian archipelago at the north west, her canoe party make their way down the chain of islands, and on each Pele burrows holes into the earth with her digging stick Pāoa, searching for a suitable home, and leaving volcanic craters as she moves. As they proceed, Pele's companions are left or disembark, populating the whole archipelago with her kin, until she reaches Hawai'i Island where she makes a home in the crater of Halema'uma'u on the volcano Kīlauea.

On the beach with her sisters one day, Pele falls asleep. Her spirit hears the sound of drumming and follows it across the archipelago to the island of Kaua'i where she meets the handsome and athletic chief Lohi'au. Pele garlands herself with fragrant flowers and presents herself as a young beautiful woman. She and Lohi'au spend many days locked in amorous fascination. Finally Pele remembers her crater, and realises she must return. She tells Lohi'au that someone will be sent to fetch him. She was not to know that Lohi'au would be so distraught after she left that he died of heartbreak.

On her return to Hawai'i Island, Pele asks each of her sisters to undertake the long and difficult journey to Kaua'i to fetch her lover back to her. Knowing her to be fickle and tempestuous, all refuse but her younger sister Hi'iaka, who agrees to the journey, so long as Pele will look after her lehua groves and her beloved companion Hōpoe. Pele agrees, and arms her sister for the journey with a smiting hand, a critical eye and a lightning skirt. She also reminds her sister that Lohiʻau is *kapu* – off limits.

Hiʻiaka survives many hazards – including sharks and dragons – on her journey, and her long adventures provide many opportunities to practise her growing skills of healing. Finally she and her entourage reach Kauaʻi, where Hiʻiaka learns that Lohiʻau is dead. Her healing powers are by now so strong that she is able to bring him back to life.

As they start the long return, Hiʻiaka learns that Pele has broken her promise, and that her lehua groves have been burned in torrents of lava, and Hōpoe is dead. She chooses to defy Pele in return, and when she and Lohiʻau reach the rim of the crater, she adorns his body and embraces him. Pele is outraged, and orders Lohiʻau to be killed in hot lava. Hiʻiaka starts to dig into the earth, with the aim of flooding the bowl of the volcano with water, extinguishing Pele's fire.

One of Lohiʻau's friends comes to avenge his second death, but instead of killing Pele, he finds himself charmed by her. Through him she learns of the perils and setbacks that Hiʻiaka faced on her journey: that her sister had brought her lover back from the dead, and had remained faithful throughout. Pele's rage subsides, and her powerful sister is invited back to her side. Harmony between them brings balance.[vi]

Rather like hair trimmed from a lover, gone dead and dull when kept in a drawer, Pele's hair looks sad and lifeless stuff behind the glass of a museum cabinet. Flea brown in the artificial light, it's hard to imagine it as the product of fire and all that passion. Its loss of lustre lends credence to a superstition attached to the stones of Hawaiʻi: that removed from the islands they carry Pele's curse. Rock-robbing tourists have taken the curse to heart and thousands of packages containing sand, stones and volcanic glass are posted back to the islands every year, many simply addressed to Queen Pele.[vii]

#09

SARSEN

Walking amid silent, long-shadowed figures making their way uphill to the Neolithic tomb known as West Kennet Long Barrow feels like a pilgrimage. From the village of Avebury in Wiltshire, the path to the barrow faces south into the low winter sun, throwing fellow travellers into silhouette. It is hard to discern particulars of dress or person: robbed of the markers of time, we're just a few more bodies in the flow of people to this site over 5,500 years.

At the top, great sarsen stones marking the entrance to the barrow leap into view. They were hauled up here, dragged on rolling poles and tipped into deep slots in the earth that wedged them in position, closing off the burial site in 2,000 BCE, more than a thousand years after it was first built. Today they act as a sheltering gateway to the excavated chambers – five lobed spaces at the east of the barrow that once held human remains, whole and in parts, deposited in two bursts of activity centuries apart. Sarsens were also used to construct the burial chamber in 3,670 BCE, originally a simple stone structure on the crest of the hill. Archaeologists found the remains of thirty-six people from 3,670 BCE: interred as bodies, their skeletal skulls and leg bones were later removed, perhaps for veneration as ancestral objects.[i]

That first phase was short – perhaps thirty years – after which the chambers were closed off with large stones, and left for over a hundred years.[ii] The longer second phase left a jumble of human and animal remains, flints, beads and ceramic vessels, which progressively filled the central chamber over many centuries. Again, the dead were placed in the chamber whole, and their skeletal remains later rearranged.

This was when the mound that covers the chambers and the Long Barrow was built. Chalk was gouged with deer antler picks from a surrounding ditch and packed together in a 100-metre-long mound. Clad in raw chalk, the white lozenge of the barrow would have been like a beacon, visible from a great distance in the rolling Downs landscape. Today it's a low grass-covered form protruding from the hilltop – were it not for the sarsens guarding the entrance, it would be just another of the Neolithic burial mounds dotting the British countryside.

West Kennet Long Barrow is the oldest surviving structure in a complex of monuments around Avebury in Wiltshire, a landscape that became a major focus for settlement, tomb building and ritual gatherings for over 1,500 years, between the Early Neolithic and the Early Bronze Age.[iii] The village itself occupies an ancient bank and ditch structure that would have been nine metres deep when first dug. Three stone circles sit within its 1.3 kilometre circumference. The outermost ring is not only the world's largest prehistoric stone circle: it is also probably the only one housing a functioning pub. Those raised on 1970s TV will remember this as the fictional village of Milbury, setting for the terrifying series *Children of the Stone*. The weathered surfaces of the standing stones, pitted with root holes and painted in yellow, black and silver-grey lichens, still suggest fantastical forms and faces.

Hundreds more sarsens form evenly spaced avenues leading from the stone circles to ancient sites of ritual importance. One avenue ends at a wooden henge and burial ground known as the Sanctuary. There is an uncanny forty-metre cone, the largest constructed prehistoric mound in Europe – Silbury Hill, built from half a million tonnes of chalk – which looks across the River Kennet to the Long Barrow, and would have likewise formed a stark white marker in the landscape. Avebury and the Kennet Valley were the site of continuous ritual use and monument building between 3700 and 1600 BCE.[iv]

Not all the sarsens have survived. Some were broken up as building stone. A number were buried during the thirteenth and fourteenth centuries – perhaps because they became associated with paganism or devil worship. Most of the stones now standing at Avebury were found and restored to their original positions in the 1930s by local marmalade millionaire Alexander Keiller. One of the stones took its would-be concealer down with it: Keiller found the remains of a body with scissors and early fourteenth-century coins crushed beneath one of the megaliths.

Sarsen is sandstone, tough as granite, which formed from sandy silt in tropical wetlands thirty-five million years ago, overlying the chalk and flint formation of the Downs. This relatively thin upper layer broke up as it firmed, and was later fragmented by glacial action during the ice ages, leaving drifts of hard stone scattered across the landscape. Sarsen is probably from Anglo Saxon, albeit a little mangled. 'Sar' was painful or troubling (a sense used by J. R. R. Tolkien for the character Saruman[v]) and 'stan' a stone, Nordic brethren to the Swedish *sten* and Norwegian *stein*.

The Downs were long notorious for their drifts of large stones: they gave the area an otherworldly character that contributed to its status as a site of peculiar power. Diarist Samuel Pepys, travelling between Stonehenge and Avebury in 1668 noted 'it was prodigious to see how full the Downes are of great stones; and all along the vallies, stones of considerable bigness, most of them growing certainly out of the ground so thick as to cover the ground, which makes me think the less of the wonder of Stonage, for hence they might undoubtedly supply themselves with stones, as well as those at Abebury.'[vi]

As Pepys surmises, most of the sarsen stones of Stonehenge ('Stonage') thirty kilometres to the south come from this same landscape, though the West Woods overlooking the Kennet Valley were only identified as the source in 2020 (see Dolerite, p. 132). In a region peppered with sarsens like so many grey sheep, why specific stones from West Woods were chosen is unclear. Perhaps they were known to be large and shapely, making the area particularly significant for Neolithic people.[vii] Releasing a megalith from the ground and the collective action of transporting it were rituals or rites of passage in themselves.[viii]

Not only stones were plundered from these sites. Dr Robert Toope harvested 'many bushels' of human remains from the Sanctuary – and perhaps also the barrow – between 1678 and 1685 for use in medical preparations.[ix] His potion was likely an adaptation of Goddard's Drops (also known as Guttæ Goddardianæ or Anglicanæ Guttæ), a potion recommended for fainting, apoplexy and bladder stones. Developed by Oliver Cromwell's physician Dr Jonathan Goddard, its concoction was complex, but the basic recipe included 'five pounds of human skull (of a person hanged or dead of some violent death)' as well as deer horn, ivory and dried vipers.[x] As a remedy its efficacy was questionable. There were some painful deaths.

More recently, remains removed by archaeologists have been

subject to reburial or 'return to the earth' requests by Druid and pagan organisations that see the creators of the sarsen monuments as spiritual ancestors, linked to the earth and stone of their sacred sites.[xi] The various strands of contemporary paganism are fast-growing religions in the West: some 21,000 people now attend the summer solstice at Stonehenge. It is not the only sarsen structure in active use – the Druid Network describe West Kennet Long Barrow as such a 'well attended' location for drumming and other rituals that 'sometimes it can be difficult to get inside'.[xii]

The climb to the barrow really is a pilgrimage, then: a route in regular use. Walking into the burial chamber, the great sarsen to the left carries deep grooves from where axe heads were sharpened 5,000 years ago. The stone facing it, to the right, carries two emblems. One is a plain mud rosette made with the imprint of a thumb, the other is filled with clover leaves and grass seed heads, around a cluster of bright red rowan berries, all so fresh they could only have been left there a few hours before.

#10

TUFF

THE HUNDREDS OF VAST STATUES SCATTERED AROUND THE quarry of Rano Raraku are often read as evidence of a once-mighty civilisation brought down by its own excesses. The quarry is set within an old volcanic crater on the island known to its people as Rapa Nui, celebrated for its sacred ancestor statues – *moai*. To the outside world it is known rather by the prosaic moniker bestowed on 5 April 1722 by Dutch commander Jacob Roggeveen, who named the island after the date of his arrival – *Paasch* – Easter.

Geographer Jared Diamond, who used the fragile ecosystem of Rapa Nui as a cautionary tale in his book on social collapse, opens his account with a wander around Rano Raraku.[i] Hundreds of *moai*, with their formidable, angular heads and dome-bellied torsos, are scattered around the crater, some still in the process of being carved from volcanic tuff. Diamond describes the quarry resembling a factory in which workers have downed tools and walked out. He ponders the twenty-metre height of the largest monolith: 'Knowing what we do about Easter Island technology, it seems impossible that the islanders could ever have transported and erected it, and we have to wonder what megalomania possessed its carvers.'

Rapa Nui is the easternmost of the Polynesian islands, claimed, since 1888, as the territory of Chile, 3,747 kilometres to the south east. In one story of its origin, the island is discovered by the spirit of the sage Hau Maka in a dream, and is identified as 'the eighth, or last, island in the dim twilight of the rising sun'. Hau Maka lived on the island of Hiva, beset by territorial struggles as coastal land was consumed by rising tides. The king of Hiva – Hotu Matua – wished to take his kin to new land, and sent explorers out to find the island Hau Maka visited in his dream. After five weeks in a canoe laden with foodstuffs, the reconnaissance party reached the island and named it 'Te Pito O Te Kainga A Hau Maka' (The Little Piece of Land of Hau Maka). Six months later, the founding families of Hotu Matu and queen Ava Rei Pua followed in a huge double-hulled canoe, carrying plant seedlings, tubers, birds, pigs and chickens.[ii]

Sitting above young oceanic crust, not far from the East Pacific

Rise, Rapa Nui formed as part of a 2,500-kilometre-long line of underwater volcanoes called the Easter Seamount Chain.[iii] Above sea level, this is a place of gentle terrain, 510 metres at its highest point, but measured from the ocean floor the island is merely the tip of a 3,000-metre volcanic cone. It is young in geological terms: the oldest rock formed half a million years ago, and there are signs of significant volcanic activity lasting up to 11,000 to 12,000 years ago. For a stone-based culture Rapa Nui offered basalt for heavy tools and hard construction blocks, obsidian for sharp blades, and tuff and red scoria as lighter, porous materials.

Tuff is formed of fine fused particles of ash and tiny stones – lapilli – ejected by a volcano. It is soft and open, good for monumental sculpture rather than fine work. The *moai* are carved in light sandy yellow tuff: some are also crowned with huge circular topknots of pitted scoria, coloured red by oxidised iron. They were formed at the quarry by teams of carvers using basalt tools, with each body sculpted from the living rock in a recumbent position and sometimes only fully separated from the tuff beneath when complete. They were then hauled by one of the dozen or so clan groups of Rapa Nui, and erected on a towering stone platform – *ahu* – within their territory. Great slabs of basalt were used for the sides of the *ahu*, above which the *moai* stood in groups, facing inland. Near some *ahu*, archaeologists have found the remains of eyes carved in white coral with red scoria centres, which may have been placed into the statues' eye sockets.

It took an entire community to manoeuvre these colossal stone figures. In 1997, using a replica concrete *moai*, Dr Jo Anne Van Tilburg, director of the Easter Island Statue Project, led a huge team in transporting and erecting the statue using materials that would have been available many centuries earlier, before the island was deforested. They achieved the best results from lashing their ten-ton statue to a wooden sledge – face down, base first – with its head and neck slightly elevated. This was pulled along a wooden 'ladder' lubricated with water and banana-stump liquid: a slow process that required the concerted strength of about fifty people and constant repositioning of the wooden supports.[iv]

Rapa Nui, once home to tropical forests, was first populated around 1000 CE. The island was largely treeless by the time Roggeveen arrived. During their brief visit to the island, his men shot twelve Rapa Nui, the opening gesture in the sequence of violence, disease and exploitation that was to follow.[v] Subsequent European visitors (among them Captain James Cook) introduced devastating infections, including smallpox and Hansen's disease (leprosy).[vi] In the nineteenth century, thousands of Rapa Nui were kidnapped by Peruvian slavers.

The history of the island told by Diamond and others is of an ecosystem over-fished and over-farmed, which suffered devastating soil erosion thanks to the felling of the trees its inhabitants depended on for canoes, cremations, cordage and the transport of *moai*. By around 1500, the island's Giant Palm became extinct. 'The reason for Easter's unusually severe degree of deforestation', according to Diamond's (not uncontroversial) theory, isn't because the Rapa Nui were improvident, but that 'they had the misfortune to be living in one of the most fragile environments, at the highest risk for deforestation, of any Pacific people.'[vii]

It may seem counter-intuitive for a society facing terrible privations to expend energy carving huge statues, but perhaps the Rapa Nui turned to the *moai* precisely because they were believed to help soil fertility. Excavation of two *moai* 'abandoned' at Rano Raraku revealed that they had been carefully positioned. Rather than awaiting transportation out of the quarry, they were part of the composition of the site itself. The *moai* were central to the idea of fertility, and may have been carved and placed in the quarry to stimulate agricultural food production – and with good reason.

Testing soil samples from the quarry, geoarchaeologist Sarah Sherwood made a revelatory discovery. Elsewhere on the island, soil was being eroded and depleted of its nutrients. In the quarry, by contrast, the soil carried high levels of elements important for plant growth and good crop yields: the result of the constant addition of small fragments of volcanic bedrock generated by carving tuff. The quarrying process had created a perfect feedback system of water, natural fertiliser and nutrients.[viii]

Rano Raraku was a significant sacred site, a place of carving and an enclosed garden where crops were grown: sweet potato, banana, paper mulberry, taro and bottle gourd. Its superior soil may have been carried to nourish depleted parts of the island. These tuff monoliths – ancestors and symbols of fertility – were dispersing Rapa Nui's mineral resources.[ix]

#11

TURQUOISE

SOTHEBY'S AUCTION CATALOGUE FOR APRIL 1854 ADVERTISED A SALE that must have provoked swooning fits in Victorian London: 'Mexican Antiquities of the Highest Interest, Consisting of a Mask and Sacrificial Knife, and a Human Scull, inlaid with Turquoise'. Bram Hertz, the shadowy vendor, touted these mosaicked objects as the treasures of his antiquities collection: 'These three objects may be considered as unique, there is no record in any catalogue of the great public museums, that such monuments of the ancient Mexican people are in existence.'[i]

Hertz was not exaggerating. Even for our generation, jaded on CGI horror, the mask and skull are uncanny, disquieting, objects. We may no longer understand their ritual significance, but they have lost little of their mesmerising power.

Lined in deerskin, with long straps once ochre-red, the skull is covered in black and blue bands of fine mosaic tiles cut from turquoise and petrified wood. Above bared teeth, the nasal cavity has been set with red thorny oyster shell. Unblinking eyes stare from the sockets: glossy pyrite irises set in conch-shell whites. The skull is thought to represent Tezcatlipoca – god of the north and the night sky, of jaguars and obsidian – and was secured by the straps to the back. The mask is the storm god Tlāloc, a blue-faced divinity associated with fertility and water. Gummed onto a wooden base, two different shades of turquoise describe a pair of serpents that form the god's face, criss-crossing one another at the nose and ringing the eye sockets.

Made by the Mexica people of the Aztec empire, some mystery surrounds how it was the fifteenth-century mask and skull arrived in Europe, though they may have made the journey as early as the following century. In his account of the conquest of 'New Spain' the Spanish priest Bernardino de Sahagún recorded gifts – very similar in description to the mask and skull in question – presented to Hernán Cortés by the emperor, Moctezuma II.[ii]

The Aztecs valued turquoise as highly as the Spanish valued their gold. In Náhuatl the word for the blue stone is *xihuitl* and it was synonymous with the quality of preciousness, used admiringly as one might use 'golden' in English – a turquoise child, turquoise words. According to the *Codex Mendoza*, three provinces in Aztec territory had to pay tribute to the emperor in turquoise: beads and mosaic tiles were tantamount to a form of currency.[iii]

Aztecs greatly prized turquoise, but they were not the first to build their communities and trade networks around it, nor had most of the stone originated in their territory. A hydrated phosphate of copper and aluminium, turquoise can only form in the presence of non-acidic copper: there are some deposits in north-west Mexico, but far more significant belts occur in Arizona, California and New Mexico, where it was mined perhaps as early as 600 BCE.[iv]

The stone is deeply woven into foundation stories of the First Nations people of the American Southwest. The Diné (Navajo people) ascribe turquoise as the colour of the spirit realm – the second of four worlds through which the first woman, the first man and Coyote passed on their way into the world of the present.[v]

Before Diné came to the Southwest, parts of present-day Utah, Colorado, New Mexico and Arizona were home to the Ancestral Puebloan culture that emerged in the first century CE. These were the people of Chaco Canyon, a complex of vast clan houses, open spaces for sacred gatherings – or kivas – and farmlands that flourished between 900 and 1150. Ancestral Puebloan society was likely to have been matrilineal, with men moving into the dwellings of their wife's family. The numbers really started to stack up: Pueblo Bonito, the largest settlement in Chaco Canyon, is a multi-storey stone and timber dwelling that extends to about 800 rooms.

This structure allowed farming and other essentials to be managed by women, who supported one another in agricultural labour and child rearing, while men left the settlement on other business: hunting, raiding, mining or trade.[vi] Fragments of turquoise and tools for working it are found all over the Chaco Canyon complex, suggesting everyone helped process the precious blue stone. Each bead would have taken about an hour to form, polish and pierce. Of the 200,000 pieces of turquoise found at Chaco Canyon, more than a quarter were located at two fabulously wealthy burial sites in Pueblo Bonita: everyone may have made beads, but possession of turquoise, at least after death, was tightly controlled.

Chips of turquoise were also embedded as offerings in the structure of the kivas.[vii] Archaeologist Frances Mathien of the University of New Mexico sees a direct link between the forbidding aridity of the region and these early farmers' veneration of turquoise: 'Blue-green is important. You look at water and the sky, what colour do you see?'[viii]

The closest source of turquoise was in Cerrillos, New Mexico, about 200 kilometres to the east, a substantial journey. Chaco Canyon may have been a site of pilgrimage, and, as at Avebury in England, the labour of constructing clan houses, kivas and monumental approaches, a devotional activity in itself (see also: Sarsen, p. 103).[ix] Turquoise from Nevada and Colorado found its way here too suggesting a flow of people exchanging valuable materials, perhaps in return for participation in rituals in the turquoise-infused structures.

Ancestral Puebloans had a long-distance trade network extending all the way to the Gulf of Mexico and Gulf of California: turquoise from the Southwest made its way down Mesoamerica and in return came Scarlet Macaw from tropical forests; various shells used for jewellery; copper; and distinctive ceramic cylinder jars used for the mixing, foaming and ritualistic consumption of cacao, the first such to be found north of the Mexican border.[x] The Ancestral Puebloans were the first North Americans to consume chocolate, and they paid for it in turquoise.

STONES AND STORIES

CALAVERITE

CHRYSOBERYL

DIAMOND

DOLERITE

LAPIS LAZULI

MOLDAVITE

MOON ROCK

OPAL

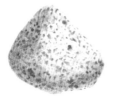

PHONOLITE
PORPHYRY

PUMICE

SPINEL

STONES AND
STORIES

F AIRY STORIES ARE FULL OF STONES. THE EVERYDAY ENVIRONMENT
furnishes our fantasies, dreaming or awake, and there is often
a rock in easy reach. In the Brothers Grimm's brutal tales, stone is
the source of both salvation and false hope: rubble is sewn into the
stomach of a goat-gobbling wolf, and bright pebbles show a shining
path in the forest of the night. In his scientific memoir *The Periodic
Table* (1975), Primo Levi reminds us that every mine is magic, its
dark recesses populated by kobolds, gnomes and goblins, supernatural being that dance in the corner of the eye.

The long tradition of creatures born from the rock beneath the
mountain laid the foundation for J. R. R. Tolkien's diabolical orcs.
'It was in fairy-stories that I first divined the potency of the words,
and the wonder of the things, such as stone, and wood, and iron;
tree and grass; house and fire; bread and wine,' he explained.[i] The
trolls of Middle-earth, petrified by daylight, join a long lineage of
crouching boulders and standing stones with legends of their own:
in Cornwall, the megaliths known as the Merry Maidens were
wicked girls who danced on the Sabbath; the rounded Moreaki
Boulders on Koekohe Beach in New Zealand are gourds that
washed ashore from an ancestral canoe.

The stony world is full of suggestion. The remains of sea creatures
found in rocks high in the mountains have inspired a remarkable
coincidence of stories across cultures. The earliest written tale of a
great deluge appears in the *Epic of Gilgamesh*, written over 4,000
years ago. The biblical story of the great flood remained in currency
until well into the nineteenth century. Found embedded in the

earth, stone tools knapped to the shape of water droplets were surely cast from the heavens by a god whose fury could be heard in the roar of thunder: 'thunderstones' are part of British and German folklore, but were also treasured in the ancient Benin Empire, placed on ancestral shrines and used for foretelling, as objects to swear by or with which to curse.[ii]

Great gems are valued for gossipy, scandalous history as well as their beauty. Jewellery collectors thrill to stones once worn by the beautiful and famous, the powerful and despotic – Mary Tudor's pearl, Elizabeth Taylor's sapphires, Marie Antoinette's diamonds – and, oh, the delectable scandal of treasures cursed, chased, stolen or concealed! As with the ruins of ancient sites, these mute stones are witnesses to history: 'if stones had eyes' we say, 'if stones could speak'.

Stones do have tales to tell if you know how to read them. The story of the Earth and beyond is still being pieced together from fragments of drama recorded in its rock. Geology is a story-telling science, requiring great leaps of poetic imagination. For geologists, the story of stone is a detective mystery in which the great question is not whodunit, but how. Starting from the present moment we are unpicking mysteries and piecing together clues, gradually working ourselves backwards through time across four and a half billion years. It is an extraordinary saga.

#01

CALAVERITE

THE AUSTRALIAN OUTBACK CITY OF KALGOORLIE TAKES ITS NAME from a vine of many edible parts, among them dark green pods with the savour of snowpeas. In English, the vine is evocatively known as silky pear. This is Wongatha country, and here the indigenous name is Karlkurla or Kurgula – pronounced gull-gurl-la.[i] With some Anglophone mangling, 130 years ago, Kurgula became Kalgoorlie.

Today the town is less associated with natural delicacies than with the adjacent Super Pit. The title is winningly literal – the Super Pit is a truly enormous hole in the ground that, until 2016, was the largest open-cut goldmine in Australia. One of the area's top attractions is the Super Pit Lookout, from which visitors can survey the enormous excavation site and, on most days, watch a controlled blast. If you're lucky, you'll visit on International Women's Day, which since 2019, Kalgoorlie Consolidated Gold Mines has honoured with a pink pit blast, with explosives arranged in the circle and cross of a Venus symbol.[ii]

The Super Pit sits on the Yilgarn Craton, a chunk of the Earth's crust thought to contain as much as 30 per cent of the world's known gold reserves. In the 120 years since European prospectors first laid claim to it, the deposit has yielded 1.7 million kilograms of gold. Until the 1980s, there were dozens of individual mines working Kalgoorlie's 'Golden Mile', and over 3,000 kilometres of underground workings, but they were all done away with, amalgamated into the Super Pit.

European prospectors first found gold in Western Australia in the 1880s, but it was Paddy Hannan and Tom Flanagan's discovery at Mount Charlotte in Wongatha country in 1893 that sparked a gold frenzy for real. The following year 25,000 men joined the rush to the goldfields. Water was scarce, and the better equipped prospectors travelled with camels brought over from India. Well suited to carrying kit over great distances in arid conditions, the camels made their way across Western Australia with their handlers from Afghanistan and territories in present-day Pakistan.[iii]

Over the next decade, 100,000 more people would follow that first rush, and the informal tent cities gradually transformed into a sizeable town. The earlier prospectors had arrived in the area without European-style infrastructure: there was a lot of building to be done, and the most readily available material for walls and roads was the stone dug up by the gold miners.

If you're on the hunt for gold, there are a few basic things you'll know before you start digging. Gold is commonly found together with quartz. It's heavy. It's soft. It's also inert – aloof, a bad mixer, one of the least reactive chemical elements. So, unlike other metals, which tend to present as ores, gold typically sits underground refusing to form compounds. In technical terminology, gold is usually mined as a pure native element.[iv]

In the back of your mind, too, you'll remember those tales of dunderheads that lost fortune and reputation digging up 'fool's gold' – crystalline rocks of brassy-yellow iron pyrite. Best leave that stuff alone unless you want to become the punchline in a barroom yarn. During the rush of 1893, prospectors found a lot of bulky pyrite-type rock, which was thrown onto waste heaps and often went on to be used as hardcore or filler in the new buildings and roads that were by then coalescing into the new town of Kalgoorlie.

Despite its standoffish reputation, gold does actually form a stable compound with the rare element tellurium, forming gold telluride – the mineral calaverite. Of the vast golden riches held in the Yilgarn Craton most – about three-quarters – is deposited as native metal. About 5 or 10 per cent of the gold is essentially invisible – present only as tiny particles or bound into the crystal structures of other minerals. A full 20 per cent occurs as calaverite.

Three years after the first gold rush at Kalgoorlie, on 29 May 1896, laboratory analysis of pyritic material found alongside the native gold leaked out to the miners: that stuff that looked like iron pyrite? – it turned out to be calaverite and worth a fortune. It precipitated a second gold rush – this one to the dumps, walls and pavements of the new town. A 1912 report in *Mining* magazine told of 'blocks of ore, assaying at 500 oz gold per ton', which had been used 'to build a rough hearth and chimney in a miner's hut'.[v] For three years, Kalgoorlie had been that mythic oddity: a prospector's town where the streets had indeed been paved with gold.

#02

CHRYSOBERYL

Among the many exercises in excess confected by Joris-Karl Huysmans for his intoxicating, decadent novel *À Rebours* (*Against Nature* or *Against The Grain*, 1884), perhaps the most notorious is an aesthetic standoff between a Persian rug and a tortoise. Allowing his eyes to meander along the silvery glints of a carpet of iridescent yellow and plum, Jean des Esseintes – an anaemic, highly-strung young duke – decides the pattern might be enhanced by a dark object moving across its surface. Purchasing a large tortoise, he submits it to the rug, but is horrified to observe the raw sepia tone of its shell saps the carpet's brilliance. 'Trying to discover a way of resolving the marital discord between these tints and preventing an absolute divorce,'[i] he decides to have the creature's shell gilded. Des Esseintes thrills to the effect, but decides this 'gigantic jewel' is only half finished, and sets to designing a floral motif for it, to be executed in precious stones.

A character of such exquisite and perverse sensibilities that he once held a funerary banquet for the demise of his own virility, Des Esseintes was never likely to be satisfied with run-of-the-mill gemstones. Not the diamond, which had 'become terribly vulgar now that every businessman wears one on his little finger.' Otherwise prized stones are likewise dismissed: rubies and emeralds remind him of the headlamps of the Paris buses: topaz, whether yellow or pink, is 'dear to people of the small shopkeeper class'; sapphire refuses to perform in artificial light. Des Esseintes turns instead to 'more startling and unusual gems', presenting his astonished lapidary with a list commencing with 'asparagus green chrysoberyls'.[ii]

A chrysoberyl, too, captivates Oscar Wilde's Dorian Gray when his journey through the beautiful, rare and exotic brings him, for a phase, to the study of jewels. 'He would often spend a whole day settling and resettling in their cases the various stones that he had collected, such as the olive-green chrysoberyl that turns red by lamplight.'[iii] Like Des Esseintes, Gray reserves these stones for private luxuriance. The lists of the two characters' favoured gems read like a knowing parody of the litany of jewels in sacred texts. In place of the Biblical – 'the sardius, topaz, and the diamond, the beryl, the onyx, and the jasper, the sapphire, the emerald, and the carbuncle'[iv] – Des Esseintes prizes the chrysoberyl, peridot, olivine, almandine, uvarovite, cat's eye, cymophane and sapphirine. For both, the preference is for gems of unsettled appearance, deceitful and mysterious.

Fit for deceitful purposes, chrysoberyl – meaning 'golden beryl' – is not a beryl at all. The pale 'asparagus green' variety is but the drab cousin to two flashier manifestations. Dorian Gray's miraculous gem, which transforms from green to red under artificial light, is a rare chrysoberyl named Alexandrite after Tsar Alexander II, who came of age in 1830 on the day the stone was discovered at the Takovaya emerald mines in the Urals. Des Esseintes's cat's eyes are also chrysoberyls, formerly known as *oculus solis* – 'eye of the sun' – for the blaze of light sparking from inclusions within the stone.[v] Mineralogists describe the cat's-eye effect rather deliciously as 'chatoyancy'.

In her essay 'Dandyism, Visuality and the "Camp Gem"', Victoria Mills notes how 'Des Esseintes and Dorian Gray reject those collections that could be found in the typical bourgeois male household (insects, stamps, coins) and prefer gender-bending accumulations of flowers, jewels, perfumes and textiles.' She suggests that both men see themselves as jewel-like objects: perfected, sparkling, refined.[vi] Chrysoberyl, with its hidden secrets, its perceptual play, its restlessness, is an apt alter ego.

Wilde revered Huysmans's work, and the 'yellow book' that poisons Dorian Gray's soul is assumed to be *À Rebours* (at the time, naughty French novels were sold in yellow wrappers). Reading the 'yellow book', Gray notes its 'curious jewelled style, vivid and obscure at once'. The rare and brilliant chrysoberyl also embodies the aesthete's commitment to a vivid life of intense sensation, epitomised by nineteenth-century essayist Walter Pater's exhortation 'to burn always with this hard, gemlike flame, to maintain this ecstasy'.[vii]

Des Esseintes's tortoise returns with its shell inlaid with chrysoberyls and the rest, and resumes its position on the rug, but the weight of its own brilliance is too much to bear, and the poor creature expires. Before *À Rebours* was published, Huysmans sent a copy to his friend Émile Zola, whose commitment to (at times grim) naturalism was quite polar to such dandy decadence. Zola's letter in reply, contained, among other critical swipes, 'A very bourgeois thought . . . It's lucky the tortoise died, because it would have crapped on the carpet.'[viii]

#03

DIAMOND

THE TITULAR GEM OF WILKIE COLLINS'S NOVEL *THE MOONSTONE* (1868) is a diamond. Huge and yellow, cool and enigmatic, like the orb in the night sky it exerts uncanny power. Described as 'the first, the longest and the best of modern English detective novels',[i] *The Moonstone* caused a sensation on publication and whetted the public appetite for tales of bewitched gems.

Collins was not the first to spin a tale of cursed jewels. As long as gemstones have been associated with magic, silver-tongued story-tellers have attributed powers both for good and ill. In his preface Collins admitted *The Moonstone* was indebted to the stories of two royal diamonds, both of which followed a troubled route from India to the palaces of Europe. The first was the Orlov, set in the Russian royal sceptre. The shape and size of a bisected hen's egg, it was a placatory gift from Prince Grigory Grigoryevich Orlov to his former lover Catherine the Great.

The second was the Koh-i-Noor, surrendered by the Maharajah of Lahore to Queen Victoria at the end of the Second Anglo-Sikh War in 1849. The enormous diamond, explained Collins, was one of the 'sacred stones of India' said to bring 'certain misfortune to the persons who should divert it from its ancient uses'.[ii]

The presentation of the Koh-i-Noor marked 250 years since Queen Elizabeth I had granted trade monopolies to the British East India Company (see also: Alunite, p. 16; Ruby, p. 54). The diamond was put on public display during the Great Exhibition of 1851, and recut according to European taste the following year. Far from marking a conclusion to hostilities, the period after the surrender of the Koh-i-Noor was bloody and unstable. In 1858, after violent revolts and reprisals, the governance of India was transferred from the East India Company to the British Crown.

Such was the context in which Collins worked on *The Moonstone*. Sixty-five years earlier, Jane Austen's *Northanger Abbey* (1803) had suggested Britain was no place for horrid intrigue: there was no space for the Gothic in a country 'where every man is surrounded by a neighbourhood of voluntary spies'. In the early years of the

British Raj, Collins instead found the corruption of Empire spreading like a stain behind the superficial politesse of Victorian society.

The moonstone of the book's title was supposedly plundered during the looting of the royal palace of Seringapatam in 1799 (a real event so violent that men of the East India Company were punished by hanging and flogging). The diamond carries its curse back to England, exposing, as it goes, a seeming wealthy elite who turn out to be living on borrowed funds, self-interested do-gooders, merciless and judgemental Christian evangelists, and entitled aristocrats blind to the feelings of the lower classes.

For millennia, India was the source of diamonds in Europe and Asia, dug from shallow mines in alluvial deposits into which they had washed from their mother rock. The earliest European reference to Indian diamonds appears in the writings of Theophrastus in 315 BCE.[iii]

Diamonds are lipophilic – they are attracted to fat and grease – an unusual quality that furnished legends of a 'Valley of the Diamonds', found in texts from China, Persia and Europe. The valley was an inaccessible crevasse into which gem hunters threw chunks of raw meat to attract eagles. The birds swooped down to pick up the meat, then carried it to the cliffs to eat, bringing the diamonds up with them stuck to the fat. The distinctive iconography of eagles dining on gem-encrusted flesh was used in texts as late as the sixteenth century to represent India.[iv]

The Koh-i-Noor and Orlov were not the only great Indian diamonds to make their way by fair means or foul to Europe. In parallel, a narrative tradition emerged of 'idols' ransacked for their (often cursed) gems.

The Regent Diamond – now in the Louvre – was acquired in 1701 by Thomas 'Diamond' Pitt, president of the East India Company in Madras. Pitt insisted he came by the diamond in honest trade, but rumours, as they tend to, proliferated. One story had the diamond smuggled by an enslaved miner out of Golconda, hidden in a wound on his leg.[v]

In 1717 Pitt's diamond was bought for King Louis XV, who carried it in his coronation crown, and subsequently wore it pinned to his hat. It later made its way into the hilt of Napoleon's sword. Pitt, meanwhile, had bought land in England with his Indian wealth: a family seat from which first his grandson and then his great grandson gained a parliamentary seat and went on to become British prime minister.[vi]

It was onto this heritage of rumour and 'colonial guilt-anxiety'[vii] that Collins constructed his tale of a yellow jewel stolen from a sacred Indian statue. The story soon inspired imitators. In 1882, a compendium of tales, *The Great Diamonds of the World*, introduced itself as 'a romance of truth' furnished with 'tragedies innumerable' including cursed gems such as the Tavernier Blue (later known as the Hope): an Indian diamond sold to the French king Louis XIV by adventuresome gem merchant Jean-Baptiste Tavernier.[viii]

It turned out to have been the perfect moment for Collins to capture the public imagination with a tale of a diamond with mysterious powers. In 1867, the year before publication of *The Moonstone*, the twenty-one-carat Eureka diamond was found in the Cape Colony, South Africa: in 1871, there were further discoveries on a farm owned by Johannes Nicolaas and Diederik Arnoldus De Beer. Soon, mines in Kimberley had produced more diamonds in a few years than India had in two millennia, making them accessible, for the first time, to the wealthy middle classes, and soon so numerous that diamond companies resorted to promotional tactics. The promise of true love was not the only weapon in their marketing arsenal: canny dealers also resorted to the macabre.

On 14 December 1910, a short article on page nine of the *New York Times* announced the arrival in town of the Hope diamond or '"Le Bijou du Roi" as it is called in France'. Readers were informed that the diamond weighed forty-four and a half carats, and that offers had already been made for it and rejected, including one of $250,000 from a man who wished to buy it for his wife as a New Year's gift. 'No other stone in existence is anything like this one,' the jeweller gave as his defence, adding conspiratorially: 'Its unusual history adds greatly to its value.'

That unusual history was recounted a month later after a sale was announced for $300,000 to Ned McLean, son of the proprietor of the *Washington Post*, and husband to the appropriately scintillating mining heiress Evelyn Walsh. The report speeds through a litany of misfortunes suffered by owners of the Indian gemstone: Jean-Baptiste Tavernier had his throat ripped out by dogs in Constantinople after selling the diamond to the French king; Louis XVI and Marie Antoinette were guillotined; Lord Francis Hope suffered catastrophes in his marriage and financial affairs. There were picturesque instances of madness, suicide, disgrace and death by misadventure.

The McLeans actually paid Pierre Cartier $180,000 for the blue stone – far less than he hoped for. Evalyn Walsh spent decades flaunting the bauble and its curse, set against lavish costumes, the collar of her Great Dane and, notoriously at one soirée, on nothing but her naked skin. She suffered the appalling loss of two children – one, aged nine, in a car accident, the other aged twenty-five, of an overdose – her husband was declared insane in his thirties, and she became addicted to opiates, adding an eventful new chapter to the diamond's legend. When she died – 'of cocaine and pneumonia'[ix] – aged sixty, the Hope diamond was bought by the jeweller Harry Winston, who, unable to sell it, toured the stone with great theatrics around the country, before donating it to the Smithsonian in 1958.

The curse of the Hope was all hooey, of course: a fiction worthy of Wilkie Collins. The first whispers of it were elaborated on to satisfy a public trained on Gothic fiction in the nineteenth century: in 1911, the story was greatly embellished by Cartier, attempting to drum up excitement for an unfashionable dark blue gem attracting little interest. Walsh's misfortunes were idle wealth, addiction and bad luck rather than a cursed gemstone, but she and others could not help but make the connection. As with *The Moonstone*, the suggestion that riches and power are founded on something dark and rotten is irresistible: the enigmatic diamond dazzling as a crystalline emblem both of magnificent wealth and of wickedness.

04

DOLERITE

THE SARSENS AND DOLERITES OF STONEHENGE IN WILTSHIRE HAVE awesome powers of suggestion. About 5,000 years ago they persuaded people equipped with only cord, and tools of stone and antler to start dragging them across the British landscape to a great open chalk plain and erect them in a grandiose configuration. They have since convinced generations to ponder, tend and preserve them. Since entering the British literary landscape alongside Merlin in the twelfth century, Stonehenge has vied with the mythic magician for dominance of the creative imagination.

The stones have been painted against moody skies by Constable and Turner, and inspired William Blake to prophetic odes. Siegfried Sassoon described the monument as an emblem of 'the roofless past;/ Man's ruinous myth; his uninterred adoring/ Of the unknown in sunrise cold and red'. More recently it has made a memorable cameo on stage with Spinal Tap[i] and has been reimagined as a bouncy castle by Jeremy Deller.[ii] Stonehenge is a lithic celebrity complete with an army of obsessive fans all ready to leap into heated disagreement over who built it, how, and why.

To Inigo Jones, viewing the site in the 1650s after a visit to Italy, the monument expressed classical architectural principles, evidently built after the Roman invasion of Britain. To the Georgian antiquarian William Stukeley, it was a sacred site of the ancient Druids, a people he imagined laying the patriarchal foundations for Christianity. In 1781, an 'Ancient Order of Druids' was reconstituted according to Stukeley's ideals. The Order has flourished since Stukeley's time, and the Druids' right to occupy Stonehenge at the summer solstice is now so firmly established that few would question its being an ancient tradition.[iii] The Druidic connection was slammed by architectural historian James Ferguson in his (less fun than it sounds) book *Rude Stone Monuments* (1872), who noted sniffily that 'had so silly a fabrication been put forward in the present day, it would probably have met with the contempt it deserved.'[iv]

Others have believed Stonehenge a monument to warriors fallen in battle, a necropolis, a calendar, a computer, a site of healing, stargazing, magic and sacrifice. It may have been – or may yet be – many of these things. Certainly in the earliest years, around 3100 BCE, it was a place of ritual burial and cremation. Archaeologist Jaquetta Hawkes's famous line that 'each generation gets the Stonehenge it deserves – or desires' reminds us not only that the unknowable early history of the monument is always open to be rewritten, but also that the stones are part of our living present. Currently they function predominantly as a site of tourism.

The latest Stonehenge disagreements relate to the smaller dolerites standing in two rings within the outer circle of towering, lintel-topped sarsens. They are known as bluestones – in the monument's early years, they would have formed a glossy dark interior within the brighter whiteness of the outer structure – and have travelled much further than their massive sandstone companions. Long before geology became a formal discipline, when alliance between stone and land was common sense, something innately alien about the bluestones was already evident.

The medieval chronicler Geoffrey of Monmouth credited Merlin with installing Stonehenge on Salisbury Plain. Geoffrey's Merlin is the miraculous son of a cloistered nun and an incubus:[v] a wise man, a prophet and an engineer rather than a magician. He is summoned by the fifth-century king Aurelius Ambrosius to erect a monument near Salisbury to honour noblemen slain in battle by the Jutish hoards of King Hengist. Merlin suggests that if Aureliu-s wishes to erect an eternal monument, he should take the stones known as Giant's Dance from Kilaraus in Ireland: 'They are mystical stones and of a medicinal value. The giants of old brought them from the farthest coast of Africa, and placed them in Ireland while they inhabited that country.'[vi] Merlin is sent to Ireland with Aurelius's brother Uther Pendragon, and an army of 15,000 men. Brute strength will not uproot the stones, but Merlin uses his 'art' and they are lifted, and at last reinstalled in the same configuration near Salisbury.

The bluestones are not from the island of Ireland but from the Mynydd Preseli (Preseli Hills) in south-west Wales, which in the fifth century of Geoffrey's tale was Irish territory. Preseli is 240 kilometres as the stone flies from Salisbury Plain. Although smaller than the sarsens, each of the bluestones weighs between two and five tons. They were probably the earliest stones erected on the site, first positioned as an outer ring in 3080 to 2950 BCE.[vii] The ritual transportation of each sarsen on its thirty-two-kilometre journey from the Marlborough Downs took about two months: the logistics of navigating a similar trek from Preseli is almost incomprehensible. Even if the stones were carried by boat, over sea and up river, it would have been a daunting journey in an era before sail-power.

Aubrey Burl, one of the twentieth century's great chroniclers of stone circles, wrote evocatively of Stonehenge, describing it as a 'ravaged colossus' that 'rests like a cage of sand-scoured ribs on the shores of eternity, its flesh forever lost'.[viii] Burl found the notion that the bluestones were transported to Stonehenge unconvincing: Neolithic circles were typically built of local stone, binding monument to place. Why, in a chalk landscape scattered with sarsens, bring dolerite hundreds of kilometres by sea or over tricky terrain? He took issue in particular with the idea that these stones had any special quality that might merit such labour, describing them as a 'mineralogical farrago'.[ix]

Burl proposes instead that bluestones arrived in the Wiltshire landscape as glacial erratics during a distant ice age, long enough ago for almost all of them to have crumbled away, leaving only the eighty that made their way into Stonehenge.

Sometimes, dead men do tell tales: isotopic analysis of the cremation burials at Stonehenge suggest that while most of the individuals were native to the chalk landscape around Salisbury Plain, a significant proportion had spent decades among the rocks of south-west Wales. The story imparted by these ancient remains, then, is of Neolithic migrants from Wales, buried near a stone monument, elements of which seem to have accompanied the human bodies on their journey. Salisbury Plain was a place of ceremonial gatherings (see: Sarsen, p. 103): this was the site to which these people apparently brought the great stones of their ancestral homeland, creating a potent blend of associations.[x]

Recent testing at Waun Mawn, a Neolithic site near Preseli, has revealed stoneholes and evidence of ritual burials and other human activity dating from 3400 to 3200 BCE. This would have been one of the first stone circles erected in the British Isles, pre-dating Stonehenge, and making it a possible origin site for the bluestones, which in their first position on Salisbury Plain, stood in a circle of similar dimensions – about 100 metres in diameter – and with the same alignment to the midsummer solstice sunrise.

The question remains as to why – was this a unifying gesture binding the people of southern Britain? Was it a conscious blending of territories? Notwithstanding Burl's dismissal of them as a 'mineralogical farrago', was there something so particular about these stones that they persuaded a Neolithic community to transport them at enormous human cost? A recent project has approached the rocky landscape of Preseli and Stonehenge with 'stone age eyes and ears' and come to the striking conclusion that some of the bluestones have pronounced acoustic qualities.[xi] There are clues to these singing stones in place names: Maenclochog, Welsh for 'ringing' or 'bell stones', is a local village that reportedly had a pair of bluestones acting as church bells up until the eighteenth century. The local quarry was called 'Bellstone'.

Hitting the rocks with hammer stones, some (far from all) will ring. The struck stones sound much like metal percussion, with tones ranging from the clarity of a bell, to the rattle of a drum, to gong-like vibrations: resonance that cuts through clear air to be heard across distance. These lithophone rocks may have invited thoughts of magic, spirits or special powers. Their acoustic qualities would also have made the process of working and dressing the stone for installation in the circle a resonant, rhythmic activity, powerfully felt by those involved. While the outer stones remain rough, the inner dolerites at Stonehenge are well worked into tall, thin pillars that to some, suggest a human torso or an axe-blade.[xii]

Their resonance struck Thomas Hardy. For the dramatic conclusion of *Tess of the d'Urbevilles*, he places Tess Durbyfield and Angel Clare in the midst of the monument on a moonless, windy night, with Stonehenge revealed to them only by sound and touch: 'What monstrous place is this?' asks Angel. 'It hums,' replies Tess, 'Hearken!' 'He listened. The wind, playing upon the edifice, produced a booming tune, like the note of some gigantic one-stringed harp.'[xiii] To Hardy, writing in the late nineteenth century, the stones were a place of pagan sacrifice: an ominous location for his doomed heroine. Today, they rest on the brink of a new, more musical, role.

#05

LAPIS
LAZULI

O F THE MANY SPLENDOURS DESCRIBED IN THE 4,000-YEAR-OLD *Epic of Gilgamesh*, the blue stone lapis lazuli holds special status. It is the material of the sacred, the royal, and even of the story itself. The introduction invites us to:

> *Find the copper tablet box,*
> *Open its lock of bronze,*
> *Undo the fastening of its secret opening.*
> *Take and read out from the lapis lazuli tablet*
> *How Gilgamesh went through every hardship*[i]

Behold the great city of Uruk, where the demi-god king Gilgamesh has become an arrogant despot, strutting his power like a wild bull, demanding warriors' daughters and young men's brides. His abused people complain to the gods, who restore balance through the creation of the mighty Enkidu, a savage with matted hair, born into the wilderness where he runs and drinks with the gazelle. To tame him, Gilgamesh sends the irresistible Shamhat, a sacred prostitute from the temple. Shamhat and Enkidu make love for six days and seven nights, transforming him from beast to man – when he is done, the gazelle keep their distance. Shamhat teaches him the ways of Uruk, and Enkidu enters the city with her as a fine, well-groomed, beer-drinking specimen of manhood. He and Gilgamesh fight, then embrace.

As companions, their adventures start with a long journey to the Cedar Forest to kill the monster Humbaba the Terrible. He is beheaded and disembowelled, and his mightiest tree felled to furnish Gilgamesh with a door. Back in Uruk, the goddess Ishtar watches Gilgamesh as he washes and dresses after the fight. Enjoying what she sees, Ishtar demands he grant her his lusciousness, offering him marriage and a 'chariot of lapis lazuli and gold'.

The goddess of love, sex and war, things have not ended well for mortals on the receiving end of Ishtar's passion, as Gilgamesh reminds her, listing the grim fates of her past lovers before rejecting her invitation. Humiliated, the wrathful Ishtar begs her father to unleash the Bull of Heaven on Gilgamesh in revenge. It is a formidable monster, which brings seven years of famine to Uruk, and blasts holes so large that hundreds fall to their deaths. Enkidu leaps into one to grapple the bull, which spews spittle and dung at him. Together the friends conquer the bull and dismember it. Ishtar descends to the city walls to curse them, but is dispatched when Enkidu throws the bull's hindquarters at her. Gilgamesh gathers the craftspeople of Uruk to study the bull's horns: 'each fashioned from 30 minas of lapis lazuli/ Two fingers thick is their casing.' Gilgamesh takes the bull's lapis horns and hangs them on his wall.

The reign of the legendary Gilgamesh is placed around 2700 BCE. His *Epic* – its themes of harmony, kinship and duty so fundamental to Mesopotamian cultures that it was imagined engraved on lapis lazuli – is of course a fable, but it is one embedded in very real material culture. Forgotten for millennia, a version of the poem written in Akkadian was found on clay tablets from the library of King Ashurbanipal.[ii] Unearthed during excavations at the ancient city of Nineveh by Assyrian archaeologist Hormuzd Rassam,[iii] they were sent to the British Museum in London and caused a sensation when translated in the 1870s.[iv]

Once the centre of an empire that stretched from the Persian Gulf to the mountains of Anatolia, and extended down the flood plains of the Nile, the remains of Nineveh sit across the Tigris from Mosul in present-day Iraq. Rassam started his archaeological career as an assistant to the gun-slinging British adventurer Austen Henry Layard. Together, they oversaw the excavation of the royal city of Nimrud, thirty kilometres south of Mosul, and uncovered the great lamassu – human-headed winged bulls that stood at the gates of Nineveh.

For centuries these cities had existed as little more than legend. The excavation of palaces, royal treasures and burial goods revealed sophisticated ancient cultures that had lived in the 'fertile crescent', between the rivers Tigris and Euphrates. Mesopotamia gave the world the earliest city-states, the first written language and, ultimately, the first empire. Its roots go back to the late fourth millennium BCE, through the ancient dynasties of Ur, to the time of Ashurbanipal, and beyond, to Xerxes and Darius. In March 2015, 160 years after they were rediscovered by Rassam and Layard, ancient stone sculptures at both Nineveh and Nimrud were destroyed by IS, who denounced them as false idols.[v]

Lapis lazuli is found across the sites of ancient Mesopotamia. It was used for beaded jewellery, the handles of royal weapons and engraved cylinder-shaped seals to roll across wax or clay. As facing for statuary, lapis was the pupils of eyes, eyebrows, beards and hair, both human and animal. The lapis horns of the Bull of Heaven recall a bull-headed lyre from Ur, with lapis tips to his horns and a lapis poll and beard. A marvellous statue of a goat standing to eat leaves from a bush has horns and a shaggy mane encased in carved lapis. Both were found in what is sensationally dubbed the Great Death Pit at Ur, thought to be the burial chamber of Queen Pu-abi who was interred between 2600 and 2450 BCE with a retinue of handmaidens and guards. The royal corpse was crowned with a leafy diadem of gold, lapis lazuli and carnelian.[vi]

An intense dawn blue, flecked with the gold of iron pyrites, small wonder the Mesopotamians prized lapis lazuli. What is surprising is the prevalence of the rare stone across so large a territory when lapis is geologically incapable of forming in the territories of Mesopotamia and Egypt. Like many of the materials most prized in Mesopotamia, it came from elsewhere. In *The Epic of Gilgamesh* matter is precious and the power that lies in transforming it revered. There are great curses in which the right and sacred qualities of things – garments, riches, bodies, beasts – are desecrated ('may a drunk soil your festal robe with vomit,'; 'May you never acquire anything of bright alabaster'). Every material thing broadcasts its lineage from raw matter into cultural object: the carved wooden door was part of the cedar forest; the axe-head, stone; beer, water; and the people of the great Mesopotamian city of Uruk were once as wild and animal as Enkidu.

Almost all of the lapis found across Mesopotamia had come from mines at Badakhshan, on what is now the border between Tajikistan and north-eastern Afghanistan: a trade route established perhaps as early as 3,500 BCE, which involved a 2,400-kilometre journey east, across desert and mountain.[vii]

Mesopotamia monopolised lapis lazuli from Badakhshan, and exported it to Egypt when stocks were high. Astonishingly, this was the same mine that supplied medieval Europe with lapis, the mineral source of the exquisite costly blue pigment known as ultramarine –

'beyond the sea' – stuff of the Virgin Mary's cloak and Titian's radiant skies. Visiting the mine in the 1960s the archaeologist Georgina Herrmann noted an abundance of soot, traces from the ancient method of cracking rock to extract lapis by the application of heat and then cold water.[viii]

The gods take their revenge on Gilgamesh and Enkidu for killing their great monsters, Humbaba and the Bull of Heaven. After a series of ominous dreams, Enkidu expires. Unable to accept his companion has died, Gilgamesh refuses to bury him until a maggot falls from his nose. It is proof of Enkidu's mortality, and forces Gilgamesh to confront his own. To honour his friend, he calls on the great artists of Uruk – the blacksmith, the lapidary, the coppersmith, the goldsmith, the jeweller – to create a statue of his friend. Like the gorgeous funerary figures of Ur, Gilgamesh orders that Enkidu is to be immortalised in the most magnificent materials available: his skin will be gold and his chest of lapis lazuli.[ix]

#06

MOLDAVITE

IF A MINERAL WERE EVER INVITED TO AUDITION FOR THE ROLE OF kryptonite in a Superman movie, moldavite would be a shoo-in. An otherworldly bottle-green substance, moldavite is a tektite – from the Greek *tēktos* or molten – natural glass found in baroque droplets around meteorite impact sites. Long considered of extra-terrestrial origin, tektites are thought to be splashy remains of rocks vaporised during a high-energy impact. Not all tektites share moldavite's vegetal good looks – they're more commonly black and turdy-looking, like interplanetary goat dung.

Formed during a single incident 14.6 million years ago, moldavite is rare stuff, found in small pieces resembling miniaturised gourds or blobby seaweed. It was first identified in 1786 by Dr Josef Mayer, living in what was then Moldauthein on the River Moldau – hence 'moldavite'. The tektites are found scattered across southern parts of the present-day Czech Republic and northern Austria. (In Czech, the town is called Týn nad Vltavou, the river the Vltava, and the stone vltavín.)

The substance 'kryptonite' first appeared in 1943 in the radio series *The Adventures of Superman*, introduced as surviving fragments of Superman's home planet Krypton which had a devastating effect on the man of steel (so much so that he could temporarily be voiced by a different actor, allowing the lead, Bud Collyer, precious time away from the mic).[i] The glowing, extra-terrestrial shards soon buried their way into the comic books as well, introducing an element of vulnerability to a character so godlike, he crushed dramatic tension as inevitably as he might a villainous plot. 'Samson's hair. Achilles' heel . . . Even the greatest heroes needed a weakness, or there would be no drama, no fall or redemption.'[ii]

Created in 1938 by two writers – Jerry Siegel and Joe Shuster – of European Jewish descent, the Superman mythos is laced with references to the Old Testament and diasporic angst.[iii] On page, kryptonite first appeared in 1949, under the editorship of Mortimer Weisinger – himself the son of Jewish migrants from Austria.[iv] Suddenly, for all to see, the alien hero was vulnerable to a substance that exposed him as an outsider: as revered comic book writer Grant Morrison puts it, 'The notion that radioactive fragments of Superman's birth world had become toxic to him spoke of the old country, the old ways, the threat of failure to assimilate. Superman was a naturalised American. The last thing he needed were these lethal reminders of where he'd come from.'[v]

It took a while for kryptonite to settle down into its familiar green crystalline form; even then, Weisinger's team introduced a whole range of krypto-variations. There was deadly gold, trippy red, plant-killing white.[vi] As unconvincing plot devices they became an easy target for satire.[vii]

In the reality-adjacent world of twenty-first century social media, moldavite has been exhibiting uncanny powers of its own. In September 2020, posts tagged #moldavite racked up over 280 million views on TikTok.[viii] Young psychics, witches and others of mystic inclination abound on the platform, and, in the summer of 2020, crystal-curious TikTokers were discussing moldavite as the 'holy grail' of stones: a powerfully transformative substance. 'There's a neat reason this tektite has become one of WitchTok's biggest and most long-lived trends,' explained one commentator. 'Moldavite removes blockages and obstacles on your path toward becoming your highest self—often in the most chaotic way possible.'[ix]

Users on the platform posted frequent dramatic updates on their journeys with moldavite. Psychics reported enhanced sensitivity. Freaky things occurred. Inexperienced crystal users were warned away in ominous terms: 'it *will* go rogue if you do not give it any direction.'[x] Forums abounded with questions such as 'Should I sleep with moldavite?' making the mineral sound like a bad prospect on *Love Island*. (For those curious, the answer is yes, but put in the work getting to know your moldavite first,[xi] and be open to encounters with 'fictitious people' in your dreams.[xii]) As the months passed, moldavite stories on TikTok took a darker turn, with the mineral implicated in the loss of jobs, friends, relationships and even family members. 'A clear message has been shared by those who have come into contact with it: do not buy Moldavite.'[xiii]

Nevertheless, buzz continued to stimulate huge demand, a spike in prices, a rush of fakes, and black-market activity. By January 2021, the Czech press reported illegal gangs in Bohemia devastating the landscape by digging deep pits in search of the valuable mineral.

Locals were advised not to approach illegal moldavite hunters, who could turn aggressive.[xiv]

As a metaphor, kryptonite has become hackneyed shorthand for a personal weakness, but for writers of *Superman* in the post-war period, the green crystal's true power was dramatic: a plot device that brought cliff-hanging peril to the previously indomitable man of steel. TikTok has common ground with the fast-moving world of comic books and radio dramas: popular accounts rely on dynamic content to keep the community checking in. The moldavite craze – and crisis – unfolded during the pandemic, when TikTokers were largely isolated at home. They had a captive audience, but nothing was happening. Enter a rare green crystal that brought chaotic energy, heartache, upheaval and plot twists: in moldavite, TikTok found the kryptonite it needed to keep fans tuning in.

#07

MOON
ROCK

WHAT IS THE MOON AND HOW DID IT COME TO BE? SUCH questions have preoccupied humans for millennia. Long before bouncing astronauts with strange modified rakes began scooping dust and stone from the moon's surface, almost every culture has offered its own explanation for the heavenly bodies lording it over the day and night. In some tales, the sun and moon are siblings; in others they are lovers chasing one another across the sky.

Selene, moon of the ancient Greeks, was grandchild to Gaia, mother of Earth and Ouranos, father of the sky, who were ardent lovers. The first of their children were Cyclopes and giants: Ouranos worried they would destroy him, so imprisoned them in their mother's womb. The next twelve were the Titans – Cronos, Crius, Coeus, Hyperion, Iapetus, Oceanus, Rhea, Phoebe, Themis, Theia, Tethys and Mnemosyne.

Tormented by the grown children trapped in her belly, Gaia begged her Titan offspring to intervene. When Ouranos came to lie with her again, four of her sons restrained him, symbolically holding heaven and Earth apart while Cronos castrated him with an adamantine sickle and cast his genitals into the sea. Ouranos dispatched, Theia, the Titan associated with sight, had three children with her brother Hyperion, god of heavenly light: Helios, the sun; Eos, the dawn; and Selene, the moon.

Selene was mother of dew, and by some accounts the seasons. The poet Sappho tells of her passion for the lovely Endymion, a mortal shepherd placed in an eternal, ageless sleep in a cave on Mount Latmos. She lives on in the science of the moon: selenology.

Before the first of the Apollo moon landings in 1969, three hypotheses were in circulation. In 1878 the astronomer George Howard Darwin (son of Charles) suggested the moon was a huge glob of the early molten Earth flung out into orbit. The second theory had the moon as a mini planet that came too close and was looped into orbit around Earth. Version three has the moon as a conglomeration of Earth's leftovers – the debris left whizzing around after our planet formed.[i]

While no hypothesis for lunar origin is provided in the universe of Wallace and Gromit, there is at least assurance as to the moon's composition (cheese) and inhabitants (a coin-operated robot cooker). When Nick Park first started animating *A Grand Day Out* (1989) as a student in 1982, it would be another two years before a formal consensus was reached among selenologists as to what the moon was and how it had formed.

MIT students studying under David Wones from the Apollo 12 Lunar Sample Preliminary Investigation Team still recall the excitement – and security protocols – of samples of moon rock arriving at their lab early in 1970. Everything relating to the Apollo missions was under tight government protection. After taking part in intensive early tests on materials gathered during the second moon landing, Wones arrived back at MIT, accompanied by moon rock samples and two armed federal agents. 'Rock' is perhaps misleading here. These weren't boulders, but regolith: loose dust and fragments from the moon's surface.[ii]

Wones and others analysing the lunar samples – igneous rock, some metamorphosed by meteorite bombardment – soon turned up anomalies that took the three existing origin theories out of play. The moon is much less dense than Earth and lacks our planet's massive iron core. The most volatile elements are absent – notably those that keep Earth covered in liquid water – suggesting they had been blasted or baked off. Further analysis indicated that Earth and the moon formed at the same distance from the sun.[iii]

In 1975, the journal *Icarus* published William Hartmann and Donald Davis's radical giant impact (or 'big thwack') hypothesis. Hartmann and Davis proposed that a smaller planetary body had a glancing impact with early Earth about 4.5 billion years ago. The mini planet was smushed, and there was a powerful explosion of liquid rock. Much of the heavier material coalesced back into the Earth, but the remaining debris orbited in a cloud, clumping together and eventually forming a moon – the moon – about 25,000 kilometres away.[iv] That distance is increasing: over the last 4.5 billion years, the moon and Earth have moved progressively further apart. Currently the distance to the moon is 384,400 kilometres and it grows by 3.78 centimetres each year.

For nine years the big thwack was just one more available hypothesis, but, in October 1984, leading selenologists gathered in

Hawai'i for a conference on the origin of the moon. Those attending remember the event as revolutionary, with the existing ideas for lunar origin resolutely discarded in favour of Hartmann and Davis's giant impact hypothesis.[v]

In the tradition of violent lunar origin stories, selenologists named the fiery, destructive, and ultimately doomed protoplanet that collided with early Earth in honour of their mythic mother, the Titan Theia.

OPAL

IN 1930 EMIL OTTO HOPPÉ, PHOTOGRAPHER OF BRILLIANTINED socialites and velvet-collared poets, left his London studio for ten months, travelling around Australia with his camera. He roved widely, from spotless middle-class homes, to Church missions where Aboriginal people were dispatched for 'protection'. In his sprawling portrait of Depression-era Australia – published as *The Fifth Continent* (1931) – Hoppé lingers over dry, demanding, landscapes, and the weather-toughened faces of people eking out a life within them.

The Coober Pedy Opal Fields were on Hoppé's itinerary of extreme locations. In his photographs, miners lounge outside the local bank and post office – rough caverns dug into the rocky hillside with corrugated sheets for roofing – and Hoppé poses at the entrance to an opal mine which looks much like the bank. One curious image shows an impromptu meat market – a lamb carcass hanging within a ramshackle wire-netting frame, a dust-covered truck from which men have leapt to purchase supplies, a counter made out of bent corrugated iron, and seated beside it, an urbane female figure in neat shoes and stockings, a fur-trimmed coat and cloche hat. A fashionable young woman in this rough place fascinated Hoppé: we meet her again in a portrait, hatless now, hair flying, light dancing in her eyes. She is identified as 'Minnie Berrington, the only woman at the Coober Pedy Opal Fields'.

Five years earlier, Minnie Berrington, like Hoppé, had left London for the antipodes. She was twenty-eight, and travelling under the Assisted Passage Scheme, leaving a dull life as a secretarial typist for the promise of sunshine and opportunity. Happenstance, vim and a fascination with opals carried her to Coober Pedy in South Australia. Travelling with her brother Victor, who drove a truck as a bush trader, she listened to men's tales of an opal field worked by independent miners with the right to keep anything they found. She wished she were a man and able to dig for opals too.[i]

With her brother and his colleague, Berrington made for Coober Pedy, where at the same troglodytic post office photographed by

Hoppé, she watched the assistant postmaster pour out a collection of fiery gemstones and felt herself 'goaded into a near frenzy to dig for opal'.[ii] Life in Coober Pedy was basic: 850 kilometres from Adelaide, with scant water, no proper roads, no built infrastructure. In summer the temperature rose to over 40 °C. The miners lived in huts dug into the ground to keep cool, but there were still the rodents and the insects to contend with. Undeterred, the Berringtons obtained miners' rights, and started digging, first excavating a fresh hole, then trying their luck in shafts abandoned by other miners.

Life for an independent miner is tough – hard physical labour, with unpredictable yields. Shafts were sunk manually, up to ten metres deep, the dry ground worked with pickaxes and shovels, and dirt – 'mullock' – hauled to the top on primitive windlasses in hide buckets. Miners search for 'potch', worthless common opal, which they then follow along its seam hoping it will lead to something precious. Minnie Berrington stuck at it for years, remaining in Coober Pedy long after her brother, and appearing as a character (the most successful of the 'opal-mining troglodytes') in Ernestine Hill's travelogue *The Great Loneliness* (1935). She dug for opal in the mornings, before the worst of the heat: in the afternoons she managed the village shop. At the end of 1933, as the summer heat was starting to peak, she heard rumours of a new opal field in Andamooka, on the lands of the Kuyani people. It was some 500 kilometres away, along unpaved and often imperceptible desert track, but Minnie had not struck lucky in months, and was eager to test her fortune in new territory.

Improbably, the gemstones of these arid places derive their fiery opalescence from water deep within their structure. Formed inside sedimentary deposits laid down in the Cretaceous, between 145 and 66 million years ago (see also: Chalk, p. 182), opals are composed of spheres of silica eroded from sandstone, carried downwards by water through faults and cracks until they reached an impermeable layer where they pool and lithify. Sometimes the faults invaded by the silica-rich water are those formed by plant or animal remains, which are slowly transformed into opalescent fossils. In 1987, a miner at Coober Pedy found the fully opalised remains of a pliosaur dubbed Eric, and now on display at the Australian Museum in Sydney.

Since the nineteenth century, Australia has been the world's primary source of opals. The most valuable are black opals from Lightning Ridge, their shadowy depths shot through with iridescent fire the colour of malachite and lapis lazuli. Mined since 1915, Coober Pedy remains the greatest source of opals by volume, and proudly describes itself as the world's opal capital, but the market is shifting. In 2008, Ethiopia emerged as a new source for opals of exceptional limpidity and play of colour, the qualities by which opals are judged.[iii] Idiosyncratic stones, their style and colour vary country to country, mine to mine: opals formed from the silica of volcanic ash in Wollo province, Ethiopia, are very different from those found at Yowah in Queensland.

Berrington's story doesn't have a Hollywood ending. She never chanced upon a vast opal that transformed her fortunes, but she was the first female opal miner in Andamooka, and in the years before the Second World War, she learned to prospect for opals, drive a truck, dig a well, sink a mineshaft, sharpen a pickaxe and lay explosives. Sometimes in all that digging, she discovered a seam of stones full of dancing light. Between her mining finds and part-time jobs (driver, general store assistant, police agent, postmistress and census taker) she supported herself as a single woman, and helped build a community in apparently inhospitable terrain.[iv]

In her memoirs, *Stones of Fire* (1958), Berrington wrote rapturously of swapping typing in soggy London for physical labour and hard living in the Australian desert. She recalls her first experience of dawn in the outback in breathless terms: 'A golden light suffused everything . . . the air was so clear it seemed to sparkle and the hills were as sharp-cut as the ones that looked so impossible on the stage . . . The enchantment of that golden serenity was so complete that I knew I would never willingly live in a city again.' Those years spent as an opal miner were, she feels, a 'dally in wonderland'.[v]

Behind the pale, sparkling eyes of Hoppé's portrait was a remarkably tough woman. In the 1940s, she broke from opal mining for war work, lying about her age to be enlisted in the Australian Women's Army Service.[vi] After years of digging in the heat, she lived to 103. Fiercely individual stones, burning with inner light, opals were a fitting obsession.

#09

PHONOLITE
PORPHYRY

Isolated within rolling prairie and pine forest in north-eastern Wyoming, eroded by years of wind and rain, stands a 264-metre-high phonolite porphyry formation, its huge bulk split into looming hexagonal and pentagonal columns. The structure resembles the columnar basalt of the Giant's Causeway and Fingal's Cave in the UK, but the scale is something else – the width of the largest columns of the Giant's Causeway is fifty-one centimetres – these are six times as wide.

There are many stories about this rock and how it formed. This one is based on an account given by Lame Deer on the Rosebud Sioux Indian Reservation, near Winner in South Dakota in 1969.[i]

Long ago, two boys ended up lost on the prairie. They had first wandered a little way from their village playing ball, then went further shooting their toy bows into the sagebrush, then became distracted following a stream lined with sparkling pebbles. They climbed a hill to see what was on the other side, then, spotting a herd of antelope, lost themselves in the excitement of tracking. By the time they realised they were hungry and tired, and started thinking of home, they no longer knew which direction home was. That night they slept curled up together beneath a tree.

For three days they walked west, eating wild turnips and berries, heading ever further from their village. On the fourth day they had the uncanny sense that they were being followed. They looked behind them, and far off in the distance saw Mato, the great grizzly bear. He was so vast that he could have plucked up and eaten the boys like berries. They could feel the Earth tremble as he gathered speed towards them. The boys ran, but the bear ran faster. His mouth opened wide, and they could see his enormous teeth set in his carmine jaws. Mato was so close the boys could smell his breath, hot and evil, when they stumbled. Though young, they had learned to pray, so they called on Wakan Tanka, the great creator: 'Tunkashila, Grandfather, have pity, save us!'

The Earth beneath began to shake, and rose up, until the boys found themselves at the top of a cone of rock. Though they were

250 metres above the ground, Mato the great grizzly could still see them, but they were just out of reach. He dug his claws in to climb up the tower of rock, but there was nowhere to catch a grip. Around and around the tower he went, leaping for the boys and digging his claws into the smooth rock. At last, when Mato had gouged the rock from top to bottom with his huge claws, he slunk away, tired and defeated.

'Of course Devil's Tower is a white man's name,' Lame Deer adds. 'We have no devil in our beliefs and got along well all these many centuries without him. You people invented the devil and, as far as I am concerned, you can keep him. Most tribes call it Bear Rock.'[ii]

The diabolical moniker came courtesy of Colonel Richard Irving Dodge and the team that travelled with him in 1875 following a tip that gold might be found in the Black Hills. The expedition translator gave the rock's name as 'Bad God's Tower', which the team then 'corrected' to devil. 'Bad God's' in itself may have been a translation mishap: the Lakota word for bad god or evil spirit is *wakansica* and the word for black bear *wahanksica*.[iii] When Dodge's team registered it as Devil's Tower, the U.S. Board on Geographic Names, which took a strong stance on the use of possessives, removed the apostrophe. So in 1906 Devils Tower – no apostrophe, no mention of a bear – became the first site designated a US national monument.

There's another famous story about this rock formation. It's about parenthood and child-like curiosity, obsession, paranoia and government secrecy. In this story, ordinary people see strange lights and flying machines, but can't convince the people they love to believe them. They find themselves haunted by a vision of the rock, compulsively drawing it and sculpting it – out of shaving foam, soil, mashed potato, rubbish and torn-up shrubs. Drawn to Wyoming, they find themselves at a secret convening of military personnel, scientists and alien lifeforms.

That tale is also the work of a great storyteller, whose name, aptly enough, means 'play mountain' – Spielberg. It's kismet that this rock formation became synonymous with the movie *Close Encounters of the Third Kind* (1977). *Jaws* had just come out, and director Steven Spielberg was busy promoting it, so production designer Joe Alves hit the road scouting for locations alone.

The script called for a strange mountain or other unusual piece of topography: What he found was Devils Tower.[iv]

How the rock was formed is a whole other story. Phonolite porphyry is an exotic material: a great chunk of igneous rock standing surrounded by sedimentary formations. Those fluted columns are the result of stress cracks that formed as liquid magma cooled. The old theory was that it was the funnel of an ancient volcano, the stone within formed fifty million years ago, then left standing alone after everything around it eroded away. More recently, geologists have suggested Devils Tower is all that remains of a lake of lava. An intrusion of magma, pushing up through many layers of sedimentary rock, hit groundwater beneath the Earth's surface, which caused a violent explosion of steam. Lava followed the passage of the steam into the resulting crater to form a lake, then cooled and solidified.[v]

Geologists use a lovely word – xenoliths – to describe the minerals pulled up from deep in the Earth's mantle when magma intrudes like this. The rock that forms afterwards is studded with crystals and fragments not usually found on the Earth's surface. Xenolith means 'alien rock' – material that is found far from its natural home. While there may not be a devil at Devils Tower, there are certainly aliens, albeit ones from the subterranean realm, rather than outer space.

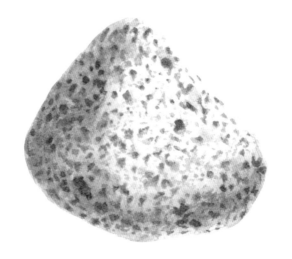

#10

PUMICE

Y OU HAVE BEEN SHARING YOUR BATH WITH A VERY DANGEROUS material. Pumice, the bubbly lighter-than-water rock used to attack hard skin on your feet, was once a foaming mass of gas and lava. Like an agitated champagne bottle, magma can carry great quantities of dissolved gases. They come out of solution violently once pressure lowers. Like a cork being popped, that pressure drops as the magma approaches the surface of the Earth, causing the release of a huge rush of gas around the vent of the volcano. Lava explodes out as molten froth, solidifying into pumice as it flies through the air.[i]

Explosive eruptions can spew out such quantities of material at speed that they cause craters to collapse. Burning hot clouds of pumice, gas and ash cascade across the surrounding landscape in a pyroclastic flow (a phenomenon helpfully explained by the rapper Ice Cube in the track 'What is a Pyroclastic Flow?' – it's what happens when volcanoes blow[ii]). Borne like a hovercraft on a low-friction cloud of gas, a pyroclastic flow can reach speeds of over 100 kilometres per hour, and temperatures as high as 700°C. The largest volcanic eruption in recorded history – at Mount Tambora, Indonesia, in 1815 – spewed a pyroclastic flow that continued to travel forty kilometres out across the Flores Sea, causing the water to boil.[iii] 'Pyroclastic' is a coinage from a Greek root meaning fire and fragmented rocks. If the flow glows, it's known by the French term *nuée ardente* – burning cloud.

Not all volcanoes behave this way. Effusive eruptions caused by less viscous magma with a lower gas content instead produce lava flows like those that created Hawai'i. Volcanologists use an index grading eruptions from the most explosive to the most effusive (a term that sounds wonderfully as though the volcano is trying to ingratiate itself).

The most powerful explosive eruptions are categorised as 'Plinian'[iv] after Gaius Plinius Caecilius Secundus – the Roman magistrate known as Pliny the Younger – whose description of Mount Vesuvius in 79 CE is our earliest detailed account of a volcano in action. From the port of Misenum on the Bay of Naples, Pliny observed a cloud of unusual size 'more like an umbrella pine than any other tree, because it rose high up in a kind of trunk and then divided into branches'. The cloud sometimes 'looked light coloured, sometimes . . . mottled and dirty with the earth and ash it had carried up'. His uncle, the natural philosopher known as Pliny the Elder, ordered a boat so that he could study the phenomenon at close hand. The trip quickly turned into a rescue attempt.

The elder Pliny captained galleys in the direction of the flaming eruption and inhabitants whose only hope of escape was now by water. 'The ash already falling became hotter and thicker as the ships approached the coast and it was soon superseded by pumice and blackened burnt stones shattered by the fire.' Disembarking, he calmed his friends and went to bed, but was woken as the courtyard rapidly filled with hot ash and the buildings shook. 'Outside, there was the danger from the falling pumice, although it was only light and porous. After weighing up the risks, they . . . tied pillows over their heads with cloths for protection.' Shortly after walking down to the shore the elder Pliny collapsed and died, asphyxiated, his nephew presumed, by heat and sulphuric fumes.[v]

Death came faster to those cities south east of the volcano, as winds showered them with hot ash and pumice for eighteen hours. Herculaneum had been destroyed in a *nuée ardente* the previous evening.[vi] Vesuvius produced a sequence of six pyroclastic surges, of increasing power. The fourth reached the city of Pompeii, covering it in hot ash. The inhabitants were subjected to a momentary heat surge in excess of 250 °C – hot enough to melt silverware but not glass – causing instant death.[vii]

Like obsidian and Pele's hair, pumice is a kind of glass, formed from silica-rich lavas. Research at Pompeii led by Sissel Tolaas is exploring what the pumice might reveal about unseen aspects of the eruption, treating the foamy rocks as time capsules – a moment frozen in stone. Tolaas is, among other things, a scientist of smell. She has previously worked with MIT to identifying the smell of fear in men's sweat, and developed an odour for a women's organisation in India, which is used to deter sexual assault. 'Stone also has memory,' says Tolaas. 'There are a lot of smells that are unique to this site, and if you don't preserve them, one way or another, soon there will be none at all.'[viii]

Traces remain of the gas trapped during the rapid hardening of frothing magma. Studying the air pockets, Tolaas has also found deposits of powdery ash captured in the moment of the explosion and preserved for almost 2,000 years. Analysing molecules harvested from the pumice, she has found traces of organic material, but most of the compounds she has discovered she has never smelled before. They do not correspond to any of the 10,000 smells in her library, which includes samples gathered at mountains in Iceland, Norway and South Africa. It is, she thinks, not a smell from the surface of the Earth, but from somewhere deeper: 'It's like the interior of the earth breathing. One nanosecond of that breath, captured in a bubble.' It's something no human would ordinarily be able to smell and survive: Vesuvius, erupting.

#11

SPINEL

POOR SPINEL. IT HAS BEEN CONDEMNED TO THE PUNCHLINE OF A joke, the impostor that is unmasked. Its only crime is to look somewhat – but not exactly – like ruby. To those who mine and polish them, the difference is apparent in the crystals, the shape of which are quite distinct. To those who handle them, spinel and ruby, side by side, present individual qualities: the one tends to be a pure, clear pinkish red, the other deeper crimson, splitting through tints of orange and violet as it tilts.

Spinel was once prized. Joanna Hardy, who dedicates the first chapter of her book on ruby to spinel, laments that it is now dismissed as a mere curiosity, or cheap substitute for ruby. Historically, spinels were valued as highly as diamonds and rubies.[i] The finest examples came from the Kuh-i-Lal mine of Badakhshan in the Pamirs, between present-day Afghanistan and Tajikistan. It has been worked since perhaps the eighth century. Their origin in Badakhshan led to the red stones being known as balas rubies, or simply balas. Unlike 'true' rubies, spinel crystals can be large, and are often left polished but uncut. The grandest spinels to pass through the treasuries of the Mughal emperors are engraved with the names of each ruler who wore them. The third Mughal emperor, Akbar the Great (1542–1605) amassed great numbers of spinels, which he wore touching his skin as protective amulets. If a ruler wore three spinels into battle, it was thought to protect him from injury or death.[ii]

The Mughal emperors were not the only rulers to revere spinel. The 400-carat red gem that crests the imperial crown of Catherine the Great; the centrepiece of the Austrian imperial crown; the Black Prince Ruby of the British imperial state crown; the Timur Ruby presented to Queen Victoria by the British East India Company: spinels all. Nevertheless, in 1946 the New York gem dealer Max Stern delighted in valuing the 170-carat Black Prince Ruby at about $20, as though, revealed as spinel, it was nothing but a worthless fake.[iii]

'Undoubtedly "rubies" now gracing all sorts of jewellery of any vintage have a good chance of not being rubies at all,' writes Paul Desautels, of the Smithsonian Institution: the miners and gem cutters in Badakhshan may have understood the distinction, but it was not until 1783 that a European scientist (Rome de Lisle) distinguished between ruby and red spinel.[iv] To attribute the stone a value of $20 seems not only unsporting, but ignores the cachet of a twisty history. That of the Black Prince Ruby is unusually rousing, commencing in an era of reckless bloodshed and bluntly nicknamed monarchs. The stone was once the property of Muhammed VI, the Muslim king of Granada, known as *El Rey Bermejo* – The Red King – for his flaming locks. It was taken from him forcibly by King Pedro – 'the Cruel' – of Castile and León who in turn bequeathed it to the dashing Black Prince – Edward, heir to the throne of England – as thanks for his support in battle.

The stone passed down to Henry V, who wore it on a crown above his helmet during the Battle of Agincourt, where it blocked an axe blow from the duc d'Alençon. Passed down through Henry VIII to Elizabeth I, it can be seen in the 'Ditchley portrait' (c. 1592) by Marcus Gheeraerts the Younger. James I set the gem in the state crown, only for it to be sold off by Oliver Cromwell – along with other royal treasures – for £15. A heart-warming twist: at the Restoration of the Monarchy, the stone was returned (at a price) for the coronation of Charles II.

It's quite a story. And as historian (and gemmological myth-buster) Jack Ogden points out, largely confected.[v] Some aspects tally with historical accounts. Muhammed VI had a magnificent treasury of gems, including many spinels, great numbers of which he brought to a negotiation with Don Pedro in 1362 in the hope he might be persuaded to combine forces against Muhammed V. This was not a well-thought-through tactic, as noted by the contemporary chronicler Lisan al-Din ibn al-Khatib: 'He might just as well have thrown himself into the mouth of a hungry tiger thirsting for blood; for no sooner had the infidel dog [Don Pedro] cast his eye over the countless treasures which Mohammed and the chiefs . . . brought with them, than he conceived the wicked design of murdering them and appropriating their riches.'[vi]

Post-mortem riffling of Muhammed VI and his entourage turned up three spinels (*balax*) each the size of a pigeon's egg, and some 700 more in baggage carried by a pageboy. There is no suggestion in accounts of the time that any of these stones were given to the Black Prince. There follow several maybes: Don Pedro may have instead given Prince Edward a golden table set with a large spinel, or one of the many large spinels listed in inventories of English crown treasures may have been the red stone from that table. (Another excellent soubriquet: the account of the jewelled table comes from the 'Black Dog of Brocéliande' aka Bertrand du Guesclin.)

Henry V did apparently wear a crown at the Battle of Agincourt and the French knights vowed to knock it from his head – one landed a blow, and the crown broke. But the king had many crowns, and there is no record of this one carrying an enormous red stone. Elizabeth I did own a large spinel, inherited from her father, but its origin before that time is unclear, as is the provenance of the spinel sold to Charles II for his coronation. The connection between the spinel and the Black Prince was made by politician and art historian Horace Walpole in 1765, based on a painting erroneously thought to show Edward with a large red stone pinned to his cap. Walpole, in turn, was adding flourishes to chivalric legends constructed around the Black Prince many centuries after his death.[vii]

Of course, some gem dealers are swindlers. Gems are submitted to many treatments to boost tone, clarity and price: colouring, bleaching, impregnating, heating, irradiating. Ruby is terribly prone to such manipulation, but spinel won't take it. So much for being an impostor gem. It may not have been carried by the Black Prince, but spinel has its champions. In 2015 the Hope Spinel – a fifty-carat gem once owned by the European banking dynasty of Hope Diamond fame – sold at Bonham's for £962,000.

SHAPES
IN STONE

AQUAMARINE

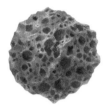

BASALT

CHALCEDONY

CHALK

GYPSUM

LINGBI

ONYX

PINK ANCASTER

QUARTZ

RED OCHRE

SHAPES
IN STONE

DO WE CARVE STONE TO ATTAIN OUR IDEAL OF BEAUTY, OR DO WE
learn beauty from stone? Carving and polishing fixes the dancing
brightness of a pebble in a sunlit stream. Eroded limestone and
sandstone formed archways long before the first temple was built.
The satin glamour of alabaster and onyx set an unattainable ideal
for human skin. (There's a reason for all those stories in which sculpture comes to life.) Stones fill us with wonder and kindle our desire.

Stone furnished the earliest artists with both medium and creative tool. Hard rocks pulverised minerals into pigments, to be mixed
with fat and body fluids to form paint. The hardest, sharpest stones
were – and still are – used to incise patterns and images, a tradition
of engraving that stretches from the earliest shapes engraved in
lumps of ochre to modern 'fantasy cut' gemstones.

We delight in the uncanny imagery revealed in stones: slices of
dendritic agate describing foggy forest landscapes; monstrous faces
hiding in swirls of jasper; marble that seems to offer a view onto a
ruined city. Intricate mineral structures reveal a shrunken world.
Among stone fanciers the most prized include those that recall their
parent landscape in microcosm: tormented rocks suggesting craggy
landscapes, miniature mountains and echoing caverns.

The great Japanese-American sculptor Isamu Noguchi immersed
himself in stone cultures, studying art that carried him back through

the earliest incisions to the 'fundamental material' itself.[i] In 1949 he took a voyage of discovery starting with the prehistoric standing stones at Avebury and Carnac, through the monuments of ancient Greece to sites in India, Nepal and Indonesia. Noguchi admired raw and tooled stone alike, working with the fine marble from Carrara and Pietrasanta ('Holy Stone') so beloved of Michelangelo, and 'fishing' for worn rocks in the Uji River in Japan.

A meditative appreciation for stone shapes is associated with philosopher poets. Since at least the Song dynasty (960–1279), evocatively formed stones were admired by Chinese literati as *gongshi* – spirit stones. In Japan connoisseurship of *suiseki* – landscape stones – relates to the artful arrangement of elements such as bonsai, in which the beauty of the natural world is distilled.

To Noguchi, stone was at once the most ancient and modern of materials: 'stone is the basis of all sculpture. Its strength is in Stonehenge. It is always new as the latest discovery on Mars.'[ii] He devoured the geological writing of John McPhee: travelogues that cross through time as well as space. Through working with stone Noguchi felt connected to Earth's deep history. In his great, glossy, carved forms and sensitive arrangements of unworked rock, Noguchi questioned the distinction between stones shaped by natural forces, and those shaped by human hands. The key consideration was not how the stone had been formed, but how you responded to it: 'Call it sculpture when it moves you so' he wrote.[iii]

#01

AQUAMARINE

AQUAMARINE IS THE SEDUCTIVE COLOUR OF HOLIDAY BROCHURES, running from noontide Aegean, to limpid Caribbean lapping white sands. It is a beryl, sister to Éire-green emerald, sun-bright heliodor and plutocrat-pink morganite.

Pure beryl is colourless. 'It turns out Earth is a pretty dirty place to grow crystals,' says Jeffrey Post, curator of Gems and Minerals at the Smithsonian, 'so natural crystals always contain some impurities.' Emerald is coloured by chromium, morganite by manganese, and aquamarine by iron. Beryls are found in pegmatites – a class of igneous rocks composed of exceptionally large crystals, formed from an accumulation of water dissolved in the magma. As this magma body cools, the part containing the most water is the last to crystallise and rises to the top, where it becomes like 'scum on a cooling pot of soup'.[i] Forget soup, this scum is the good stuff: it's going to turn into big crystals, including, if fortune favours you, aquamarine.

In the 1980s, fortune favoured three prospectors in the Brazilian state of Minas Gerais who found and extracted the world's largest aquamarine crystal – a cartoonish spike a metre long. The unwieldy forty-five-kilogram gem shattered into three sections, and was claimed by the owner of the mine. He sold the shorter pieces for gemstones, and stashed the remains of the spike, still sixty centimetres long, behind his bed.[ii]

The German town of Idar-Oberstein – a centre of stone carving since the fifteenth century – has particular links to the gem-rich region of Brazil. Transatlantic phone lines buzzed with rumour of the titanic aquamarine, and before too long, German gem dealer Jürgen Henn was standing in the mine owner's backyard, viewing the remaining piece of the crystal laid out on a table. Many huge and magnificent gems could be cut from that aquamarine, but Henn felt uncomfortable breaking this natural wonder apart. Instead, he wanted to place it in the hands of Bernd Munsteiner, a revered Idar-Oberstein lapidary and pioneer of the 'fantasy cut' – sculptural designs carved into the crystalline depths of large stones.

Negotiation for part-ownership of the gem took many years, but its destiny was laid down. It would not grace ear, neck or finger: instead the aquamarine would become that peculiar phenomenon, the *objet de virtu*. (The term is an English coinage to give luxury curios continental allure – it has nothing to do with virtue as an ethical concept.)

As it happens, another pioneer of the *objet de virtu* had a particular affinity with aquamarines. This was Carl Fabergé, jeweller to the Imperial Russian Court, master of stone figurines, fancy Easter eggs and other folderol, who also produced elegant jewels for his clients: trembling tiaras, necklaces that melt like ice, starburst brooches. 'Fabergé didn't use tremendously valuable stones: it was purely about the contribution the stone made to the design, the artistic potential,' says Kieran McCarthy, chronicler of Fabergé, and a director at antique jewellery dealers Wartski. McCarthy is on intimate terms with the Russian jeweller's original design books: 'aquamarine comes again and again – we could argue that it's Fabergé's most used stone after diamond.'[iii]

The attraction was twofold. The aquamarines used by Fabergé were Russian, part of a lapidary tradition of honouring the country's redoubtable mineral wealth through ingenious design. The Winter Palace in St Petersburg does so on a grand scale, broadcasting Russia's lithic might from its urns and columns: malachite, lazurite, porphyry, aventurine (see also: Malachite, p. 36). Almost all the stone ornaments created in the Fabergé workshop celebrate Russian minerals. There's a vase, an egg and a climbing frog in spinach-green Siberian jade; hard stone carvings inspired by Japanese *netsuke* which display a whole palette of jaspers, agates and quartzites; a Kremlin clock tower carved in rhodonite, a rosy, black-veined stone discovered in the Urals in the 1790s, after Catherine the Great pushed for development of Russia's mining operations. Why set a royal jewel with imported sapphires or emeralds when you could instead show off a Russian stone, clear as spring meltwater?

Aquamarine was also impeccably matched to the colour palette favoured by Russian high society in the late-nineteenth century. Interiors and wardrobes were awash with lilac and other muted pastel colours: from flower arrangements, to parasols, to silk dresses. An enormous Siberian aquamarine sits at the centre of a Fabergé brooch given by Tsarevich Nicholas Alexandrovich – later Emperor

Nicholas II – to his fiancée, the future empress, Alexandra Feodorovna, in August 1894, three months ahead of their marriage. The stone is from Siberia – it has a frosty, greenish clarity, like a shard of ice broken from the surface of a pond. The imperial couple carried it into exile after Nicholas II's abdication in 1917; it was one of the jewels removed from them ahead of their execution the following year.

Fabergé's designs in aquamarine evoked the stone's liquid glister: a tiara from 1904 is crested with aquamarines shaped like teardrops, as if they really were water, arrested in motion. One of the royal eggs contains a swan swimming on a little aquamarine pond.

Such whimsy was not on the cards for the giant Brazilian aquamarine entrusted to Bernd Munsteiner. The lapidary was relieved that the stone had already shattered: any faults and tensions had been released, and it would withstand carving. Munsteiner spent four months studying the stone, developing a design and planning his strategy. To keep the crystal's length, he settled on an obelisk, into the interior of which he would carve a sequence of starbursts. Cutting the design took six months – Munsteiner only worked on the stone two hours a day to keep his hands steady and his eye sharp.[iv]

Bernd Munsteiner named his design 'Ondas Maritimas' – the waves of the sea – but the stone itself was dubbed the Dom Pedro, after the (appropriately priapic) first emperor of Brazil. Thirty-six centimetres tall, and weighing almost half a kilo, the radiant blue shaft now greets visitors at the Hall of Geology, Gems, and Minerals at the Smithsonian's National Museum of Natural History. It is in excellent company: the museum looks out towards the great assertion of masculine power the Dom Pedro echoes in shape, the Washington Monument.

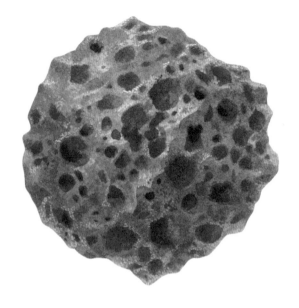

#02

BASALT

A HELICOPTER CLIMBS AND DIPS ROUND AND AROUND ROBERT Smithson's *Spiral Jetty*: the light bouncing off Utah's Great Salt Lake is now blinding silver, now garnet. Beneath the spiralling blades of the chopper, Smithson's wife Nancy Holt films her husband running centripetally around the slim basalt structure invading the lake. We draw ever closer to Smithson as he approaches the centre of the spiral, as though man, camera, helicopter, all are being sucked into a whirlpool.

In April 1970, Smithson and a local construction worker trucked 6,650 tons of black basalt from the shores into the water of the Great Salt Lake, building a spiralling walkway almost half a kilometre long. *Spiral Jetty* became the totemic work of the Land Art movement, one of the few large earthworks that Smithson completed before his death, at the age of thirty-five, when a helicopter crashed as he surveyed the site of what would be his final work, *Amarillo Ramp* (1973).

Without the film of Smithson running along the jetty, it's hard to grasp the scale of the structure. The landscape is saturated with minerals that deter larger lifeforms – there are no trees or signs of faunal life – but the lake sustains a colony of bacteria that turn the water plum pink. Were it not for the basalt spiral, we could be on early Earth.

Smithson had read of Laguna Colorada, a salt lake in Bolivia tinted by red algae and excitedly imagined working in such an otherworldly landscape. The Utah Parks Department told him that north of Lucin Cutoff – the railway bisecting Great Salt Lake – the water was the colour of tomato soup. 'That was enough of a reason to go out there and have a look.'[i] Smithson and Holt flew from New York, and drove around the lake to Rozel, where abandoned machinery and crumbling jetties evidenced an earlier struggle to exploit the crude oil seeping up through porous basalt.

'About one mile north of the oil seeps, I selected my site,' wrote Smithson. 'Irregular beds of limestone dip gently eastward, massive deposits of black basalt are broken over the peninsula, giving the region a shattered appearance . . . Under shallow pinkish water is a network of mud cracks supporting the jig-saw puzzle that composes the salt flats. As I looked at the site, it reverberated out to the horizons only to suggest an immobile cyclone while flickering light made the entire landscape appear to quake.'[ii]

Smithson was unsentimental about landscape: rather than wild, virgin territory he saw evidence of human intervention going back millennia, with all the territorial struggles and dashed hopes that carried.[iii] Widely read in geology, he saw human existence within a vast whirl of time, and the lake and stone as evidence of a cycle of change and movement. White salt crystals, pink water and black basalt provided the 'paint' for this artist's landscape, but they also offered a powerful conceptual framework. In the mineral-saturated water, Smithson saw an analogue of human blood and the ancient ocean, with the spiral returning our bodies to the primordial ooze. In basalt, he was using the raw, volcanic material of Earth.

In the same era that Smithson caused ructions in the art world, discoveries were made in the depths of the ocean that shook our understanding of how Earth formed and functioned. It was found that basalt covers two-thirds of the surface of the Earth: 'the lining of the basins that cradle the seas', writes palaeontologist Richard Fortey. 'Down, down, far beneath the waves, into the dark abyss where no light reaches except the luminous flashes of deep sea organisms, the sea floor is dressed in sediment – but this is only a thin blanket atop five kilometres or so of basalt.'[iv]

The ocean floor was only provisionally mapped and explored in the decade before Earth-dwellers landed on the moon. In the 1950s marine geophysicists Bruce Heezen and Marie Tharp, working at Columbia University in New York, analysed SONAR readings and seismic data from boats in the Atlantic. In the middle of the ocean, they found a submerged mountain range over 1,000 kilometres wide and 16,000 kilometres long, running the length of the Atlantic, sliced all the way down by a valley. Tharp suggested this was a rift – a splitting point in the Earth's crust – running down the centre of the Atlantic.[v]

Tharp and Heezen's full topographic map of the ocean floor, published in *National Geographic* in 1967, became a fixture in geography classrooms. Tharp's hypothesis of mid-ocean rifts laid the foundations for the theory of plate tectonics that emerged during the 1960s. Dredging of the Mid-Atlantic Ridge revealed that basalt is the substance fundamental to the construction and regeneration of our world. As the oceanic plates slowly separate, lava pushes up through the cracks opening along the rift: the eruptions do not push the plates apart, rather they are a consequence of movement already in progress. Older rock moves away on either side, fresh basalt forms at the rift: this, writes Fortey, is how the ocean floor grows, 'by stealth in the dark'.[vi]

During the 1950s, a piece of submarine-hunting technology called a magnetometer was adapted to read information from minute crystals of magnetite within the iron-rich basalt. The Earth's magnetic field varies slightly, year to year. As the basalt magma cools, these crystals freeze in the direction of Earth's magnetic field 'like tiny compass needles', thus preserving the orientation of Earth's magnetic field on the exact date when the rock hardened.[vii]

What oceanographers discovered was 'a bizarre magnetic pattern' that was extraordinarily regular and completely unexpected. Close to rift valleys in the Atlantic and Pacific, basalt displays normal magnetic orientation, pointing in the direction of the north pole, but a few miles to the east and west it flips by 180 degrees, indicating that the north magnetic pole was once where the south pole should have been, and vice versa. Several miles further in either direction, and the magnetic signal flips again. The pattern is repeated over and over again, in strips parallel to the rift, moving outwards, evenly to east and west. Radiometric dating of the basalt shows reversals either side of the rift happen simultaneously.[viii]

The conclusions, presented in 1961, were twofold.[ix] Firstly, that the Earth's magnetic field flips 180 degrees roughly every 500,000 years, and has done so for at least 150 million years. Secondly, that mid-ocean rifts produce new basaltic crust at a rate of a few centimetres a year. The rate of sea-floor spreading is recorded in the age of the basalt, with that nearest to the continental margins the oldest, formed more than 100 million years ago. The outer oceanic crust plunges beneath the continental shelf, returning to the mantle of the Earth: far off shore, ocean trenches form where the basalt is subducted, often accompanied by a zone of intense seismic activity.

'Art communicates not only through space, but also through time,'[x] wrote Smithson in 1966, the year before Dan McKenzie published the first paper on plate tectonics in *Nature*. *Spiral Jetty* was Smithson's answer to the puzzle of how to represent a landscape reeling 'back into the millions and millions of years of "geologic time"'.[xi] The basalt spiral echoed the structure of a growing salt crystal. It recalled the unfurling fern fossils found in South America and Africa that in 1861 led Eduard Suess to hypothesise on the existence of the ancient southern supercontinent Gondwana.[xii] With basalt now the chronometer of a shifting world, it suggested the spiralling units of measurement in geological time: eons, eras, periods, epochs and ages.[xiii]

Spiral Jetty has become part of the environment of the salt lake. It disappears for years, reemerging dressed in a carapace of salt crystals as the waters rise and fall. Smithson described his earthworks as 'abstract geology,' eliding boundaries between nature, culture and science. He described the strata of the Earth as 'a jumbled museum' and the sediment as 'text' that we must learn to read if we are to 'become conscious of geologic time'.[xiv]

Such interdisciplinary thinking was in phase with the time. The unifying theory of plate tectonics was pieced together from the concerted work of cartographers, geophysicists, oceanographers, palaeontologists, paleomagnetists, seismographers and daredevil ocean explorers. They were brought together from separate disciplines around the world by the basalt slowly forming as their respective continents spread apart.

#03

CHALCEDONY

D ESCRIBING STONES, WE OFTEN REACH FOR THE LANGUAGE OF flesh, both animal and vegetable: the finest Burmese rubies are the colour of pigeon blood; garnets have the juicy lustre of pomegranate seeds; the mineral olivine is indeed the yellowish green of Mediterranean olives. The group of silica minerals clustered under the name chalcedony – quartz hardstones composed of microscopic crystals – provide a lapidary's palette of fleshy tones and textures. There is carnelian, which can be found with the muscular, marbled texture of raw beef; chrysoprase the spring green of a Turkish plum; stripy, caramel-coloured agates; bloodstone, like a dark mound of wilted spinach flecked with red chilli; and jasper banded like *prosciutto crudo*.

Diligent stone carvers have exploited the *trompe l'œil* possibilities of chalcedony. A paperweight sculpted in Yekaterinburg for the Paris *Exposition Universelle* of 1867 celebrates Russia's wealth of semi-precious stones as a salver of summer fruits: pink and white mulberries, red grapes and a yellow plum are nestled between dark green stone leaves.[i] To one side, twigs of white and red currants have the perfect veined translucence of fresh berries. Draped on a cake, you would break a tooth before taking a second glance.

Qing dynasty hardstone carvings shown at the Metropolitan Museum of Art in New York in 2016 included a fragrant 'Buddha's Hand' citron carved in carnelian, glossy pomegranates in agate, and a dish of chalcedony peanuts and jujube dates. For each, the innate qualities of the materials add to the illusion: the dates are sticky-dark and translucent, while the coarse skin of the peanut shells is carved in a rougher opaque stone. The playfulness is not only optical: each carving is a rebus, offering a word puzzle for those receiving the gift. Buddha's Hand – *fo shou* – sounds similar to the words for 'fortune' (*fo*) and 'longevity' (*shou*), while peanuts and jujube dates rhyme with the phrase 'may you have children soon' (*zao sheng gui zi*), a common wish for newly married couples.[ii]

The grains, lines and speckles of jasper are caused by trace elements and inclusions: it's basically flint in flashy evening wear.

The *Mona Lisa* of chalcedony carvings – in its celebrity, at least – is a 5.73-centimetre-high piece of banded jasper in the collection of Taipei's National Palace Museum. Known simply as the 'Meat-Shaped Stone', the piece resembles a chunk of slow cooked belly pork, complete with dimpled fat and hair follicles. Its skin is a rich, glossy amber colour, as though glazed in soy sauce. The whole piece tilts in a top-heavy gelatinous way that suggests yielding flesh.

At the National Palace Museum the piece receives over 5 million visitors every year. Crowds are kept moving past it in procession.[iii] When it paid a two-week visit to Japan in 2014, 6,000 people crammed in to see it each day. The Meat-Shaped Stone's first trip outside Asia – for exhibition in San Francisco in 2016 – was undertaken with great fanfare. The stone generated social media frenzy (#pricelessporkbelly), and twelve chefs around the city offered braised pork dishes in tribute.[iv]

Depending on your feelings about fatty pork it is perhaps not the most poetic object, but the Meat-Shaped Stone is layered with associations. The carved and dyed jasper is a life-sized portrait of Dongpo Pork, a dish named for the Song-era poet, calligrapher and gastronome Su Dongpo, credited with inventing the dish in the eleventh century.[v] Dongpo Pork is prized for its textural balance of slow-cooked meat, soft fat and glazed skin. (Legend attached to the dish has Su Dongpo so involved in a game of Chinese chess that he forgot his cooking pork, hence its intense reduction.) For modern fans, the stone's charm lies principally in its illusion. In the nineteenth century, the jasper's uncanny resemblance to casseroled meat would also have carried literary associations: a visual allusion to Su Dongpo.

There is still a fascination with stones that look like foodstuffs, whether carved or naturally occurring. Chefs and lapidaries have created feasts of carved foods for stone fairs in China, including dishes from the legendary Manchu-Han Banquet, supposedly served by Qing dynasty emperor, Kangxi. The menu included delicacies such as 'Braised bear paw with carp tongues' and 'Imitation leopard foetus'.[vi] During the stone festival at Alashan in Mongolia, unusual specimens found in the Gobi desert are often presented on a table as if a feast of berries, tubers and fried croquets.[vii] For those whose appetites have been whetted by the Meat-Shaped Stone, here, too, are ruddy fragments of chalcedony, all fleshy and raw.

#04

CHALK

WALKING ABOVE THE VERTIGINOUS WHITE CHALK CLIFFS OF Beachy Head in the late spring, the grassland is buzzing with newly invigorated insects, hovering and darting between clusters of blue and yellow flowers. The ground is springy, warm and fragrant with wild thyme. Birdsong rings clear in the coastal hush. Here, the chalk lying so close beneath thin topsoil brings softness to foot and ear.

The great chalk formation – with its trout streams, beech woods, barrows and cliffs – is emblematic of south-east England. It once formed a land bridge to continental Europe, thus the selfsame stone is also emblematic of north-west France, extending to the Cap Blanc Nez, the cliffs of Etretat (backdrop to the Arsène Lupin adventure 'The Hollow Needle'[i]) and vineyards of Champagne. The formation extends beyond, through Belgium, to the Netherlands, Denmark and the Baltic, and gave its name to the Cretaceous, the time of chalk (*creta* in Latin), identified by the extravagantly named Belgian geologist Jean-Baptiste-Julien d'Omalius d'Halloy.[ii]

Fossil-fanciers may complain about meagre finds on the chalk, which yields plenty of flints but few ancient treasures. This is a matter of perspective. All that powdery white limestone is composed of the calcium carbonate remains of ancient plankton – foraminifera and coccolithophores – tiny, single-celled animals and algae. Over thirty-five million years, their microscopic remains settled in such abundance at the bottom of the clear, tropical seas of the late Cretaceous that in places the formation is 300 to 400 metres thick.[iii] 'So now the astonishing truth is clear,' writes palaeontologist Richard Fortey: 'The Chalk *is* fossils, virtually in its entirety.'[iv]

The Cretaceous ended with a bang sixty-six million years ago – the Chicxulub asteroid whalloped Mexico and volcanic activity engulfed the Deccan Traps in Western India – precipitating the extinction of all non-avian dinosaurs.[v] A walk across the chalk, then, is also a walk along the boundary of a momentous transition, between the era of the dinosaurs, and the rise of the mammals.

At some point, thousands of years ago, some of the mammals in southern England took it upon themselves to use the chalk as a means to depict some of the other mammals. The White Horse at Uffington in Oxfordshire has that spare, elegant dynamism that many modern artists admired in ancient art. A bounding impression, passing at speed, it was carved into the hillside 3,000 years ago, by far the oldest of England's surviving chalk geoglyphs. It has endured thanks to a centuries-old tradition of 'scouring' the White Horse. The horse is formed of deep trenches cut into the hillside, which were then packed with chalk. Every decade or so, it is scrubbed out, the edges neatened, and new chalk hammered tightly in place so that the rain doesn't carry it off in milky runnels down the hillside.[vi]

Prone to regular trims and energetic scouring, who knows how much the White Horse has been modified over the centuries? Writing in 1681, Thomas Baskerville estimated the animal's tail to be sixty yards (fifty-five metres) long, but judging the mid section 'too gaunt and slender for the length and proportion of the horse', suggested that those who are under obligation to repair the land-mark 'might do well to make the belly bigger.'[vii]

Thanks to centuries of 'improvements', and liberal garnishes of folklore and local history, the chalk geoglyphs are hard to date. The seventy-two-metre Long Man of Wilmington in Sussex, who appears to be heading for a hike with Nordic walking poles, was long assumed an ancient figure, possibly Iron Age, although the earliest record of him is an illustration made in 1710. In the nineteenth century the outline of the Long Man was marked out in yellow bricks and the position of his feet changed. Covered up during the Second World War so that enemy aviators could not use him to navigate their arrival at the south coast, in 1969 his outline was filled with painted concrete blocks.

The White Horse at Westbury also fell foul of mid-century bodging, and was filled with painted concrete in the 1950s. Local records suggest the original horse had been carved into the chalk below an Iron Age hill fort in the late seventeenth century: one of a number gouged into the English countryside in the period to commemorate historic Saxon victories over the Danes. This new tradition in turn came from the belief that the White Horse at Uffington was a Saxon battle monument, underselling the charger by almost 2,000 years.

Carved into the chalk above a pretty brick and flint village in Dorset, complete with ruined abbey and holy well, is the most charismatic and perhaps notorious of England's geoglyphs, the Cerne Giant. Fifty-five metres tall, and brandishing a knobbly club and expressive eyebrows, the Giant has a delightful raised green grassy nose, but this is seldom mentioned, so entirely is it upstaged by his eleven-metre erect penis, crisply outlined in white chalk. Theories as to his identity range from Hercules to Oliver Cromwell to the Saxon god Helith.[viii] Some local historians have assumed him to be a defiant insult, others simply celebratory, though the proximity of this lusty pagan figure to an ancient abbey has caused more than the Giant to raise his eyebrows.

Whatever his origin, the Giant's handsome state of arousal has earned him quite a reputation. Avant-garde poet, artist and all-round bohemian muse Iris Tree developed a fascination with the grassy gallant (perhaps it was the green: she once wrote a erotic poem starring a frog). In the late 1920s, she and her friend Lady Diana Cooper were both 'desperate to have a child', according to Cooper's son, historian John-Julius Norwich. His mother did the rounds of London's best doctors, but 'Iris, typically, went and lay prone on the erect penis of the Cerne Abbas Giant.' She remained out all night, and the Giant worked his magic. Norwich noted: 'Her son is 6'7" and is now a psychiatrist in Stockholm.'[ix]

In 1958, Tree's niece Virginia had trouble conceiving. Aunt Iris suggested that Virginia and her husband of five years, Henry Frederick Thynne, 6th Marquess of Bath, 'pay their respects' to the Giant.[x] A daughter was born ten months later. 'We gave Cerne as Silvy's middle name and made G. Cerne godfather at her christening,' the Marquess explained. 'I suppose it might seem strange to have a pagan figure as a godfather, but we felt he deserved some recognition.'[xi]

The Giant is less open to callers these days. His soft chalk body is vulnerable to erosion so a fence now guards his modesty and his only company are the sheep that crop his grass. Every ten years he endures a thorough manscaping, and has the old chalk dug out of his trenches and seventeen tonnes of fresh white stone hammered back in.[xii] In 2020, the National Trust celebrated one hundred years of custodianship by having state-of-the-art tests done on soil beneath his trenches. The results point to the Giant being late Saxon, looking spry at just over 1,000 years old, around the same age as the abbey, which was founded in 987 CE,[xiii] though the relationship between the two structures remains unclear. For a fifty-five-metre-tall naked man with an erection standing on a public hillside, he retains surprising mystique.

#05

GYPSUM

JEAN NOUVEL'S EXPANSIVE, EXPENSIVE, TECHNICALLY AMBITIOUS National Museum of Qatar building opened in Doha in 2019. The lauded French architect searched for an emblem that evoked art and nature, ancient and modern, and settled on a gypsum formation known as the desert rose – a natural growth of fine interlocking sandy-coloured discs which he translated at a vast scale into the building's attention-seeking structure. Architecture critics at the time likened Nouvel's outsized (reinforced concrete) crystal discs to a piece of 'accelerated geology', noting how interior spaces served the radical shape of the structure: the focus of this museum was its building, the displays more or less an afterthought.[i]

In nature, the desert rose is formed of sand-infused crystal 'petals' that grow after rain in the arid landscape: it is a piece of geological flamboyance from a mineral shape-shifter. Gypsum takes many forms. One is the clear, spiky, crystal selenite (after Selene, goddess of the moon: see also Moon Rock p. 146). The Naica mine in Chihuahua, Mexico, is home to two selenite formations that make Jean Nouvel's structure look blushingly conservative. Surfaces in the Cave of Swords, found in 1910, bristle with metre-long, blade-like crystals. If that reminds you of an Indiana Jones movie, the second cave, opened in April 2000, is pure sci-fi. The chamber is spanned by gargantuan crystals, up to twelve metres in length, which dwarf visiting scientists in their hazmat suits.

The crystals in the Naica mine have been growing for perhaps half a million years, but they're fragile, and deteriorating now that they are exposed.[ii] Gypsum is soft – 2 on the Mohs scale, just above talc – no good for jewellery, but easy to carve.[iii] In its fine-grained 'massive' form, gypsum is one of two stones known as alabaster (the other is a slightly harder calcite somewhat resembling marble). It is a soft, bright white stone used for detailed interior work such as the intricate, silken relief panels and human-headed winged bulls (lamassu) carved for the palace of Ashurnasirpal II at Nimrud in the ninth century BCE.[iv]

Gypsum is ubiquitous – there is probably some in the room with you. Roasted and pulverised, it can be reconstituted as a fast-drying paste that hardens to a satiny, soft-stone finish. Its use in interior plasters dates back to ancient Egypt: fine gypsum applied over carved limestone gives Queen Nefertiti her creamy skin.[v] The lacy, fretted designs spidering across the walls and ceiling of the Alhambra Palace in Granada were made in gypsum plaster, moulded, carved and painted. Sandwiched between two sheets of paper, gypsum is the filling for plasterboard, a labour-saving staple of flimsy modern buildings, now causing an environmental headache.[vi] Jean Nouvel's 'desert rose' not only resembles gypsum: consciously or not, the building will, to an extent, be made from gypsum too.

Refined, pulverised gypsum is plaster of Paris, mined and processed in Montmartre until the 1780s. Early in the eighteenth century, plaster was introduced as a support for broken limbs: a disastrously cumbersome block, formed by pouring the liquid around the injured arm or leg in a wooden frame. In 1839 Lafargue of St Emilion instead dipped linen strips into plaster to form the first lightweight moulded casts for fractured limbs.[vii]

Other generations have had their culture wars, and gypsum plaster has the dubious distinction of having been a victim of two of them. Plaster of Paris can be used to form both moulds and casts, allowing a three-dimensional form to be reproduced almost endlessly. In the eighteenth century, plaster casts schooled Europe in the appreciation of classical statuary. Copies of the Farnese Hercules, Apollo Belvedere, Laocoön and their marmoreal kin were 'the apostles of good taste to all the nations', wrote the philosopher Denis Diderot in 1763.[viii]

So closely was gypsum plaster associated with aesthetic refinement, that German art historian Johann Joachim Winckelmann reached for it in describing his distinctly erotic response to great sculpture. 'The true feeling for the Beautiful is like fluid gypsum that is poured over the head of Apollo and reaches and envelops every part of him,' he purrs in his voluptuously titled *Treatise on the Capacity for Sensitivity to the Beautiful in Art and the Method of Teaching*.[ix] In line with the ideas of Winckelmann and Diderot, the wealthy families of Europe amassed collections of gypsum casts, bringing the glories of Rome to Derby and Dresden.

The encyclopaedic museums opening across Europe and the US in the mid-nineteenth century acquired substantial collections, too. The Museum of Fine Arts in Boston decided it would be too costly and time-consuming to assemble a collection of original antiquities: gypsum casts were affordable and available.[x] In the early days of the Fitzwilliam Museum in Cambridge, 'casts and originals of all periods were displayed side by side, apparently indiscriminately' in an 'Aladdin's cave of *collectables*'.[xi]

The casts established a canon of great works, an ideal of beauty distinguished by the radiant white of gypsum plaster, reinforced by its repetition in the world's great collections. At the Victoria and Albert Museum in London, Henry Cole established the International Exchange Scheme for the Promotion of Art, allowing casts to be shared and circulated around Europe. They could also be purchased for private edification: the first British Museum catalogue contained a list of copies that were available and for sale.[xii] Exported to the 'New World', the reproductions disseminated a (very white) ideal of European culture.[xiii]

In the early twentieth century, casts fell from grace. In the 1920s, about 40 per cent of the Parthenon Frieze on display at the British Museum was plaster but in the following decade the casts were removed: the display would only include original material, no matter how fragmented.[xiv] Copies lacked the rarity and exclusivity that excited the competitive new generation of connoisseurs. In his 1936 essay *The Work of Art in the Age of Mechanical Reproduction*, Walter Benjamin sounded the death knell: 'even the most perfect reproduction of a work of art is lacking in one element: Its presence in time and space, its unique existence at the place where it happens to be.'

For want of the magical ingredient Benjamin described as 'aura', casts were consigned to the academies, to be drawn and studied by generations of art and archaeology students. It was here that they faced and lost their second cultural skirmish. In the latter half of the twentieth century, as revolutionary fervour zipped through art schools in Europe and the US, gypsum casts were ousted from the drawing studios too. Falling foul of the new hunger for authenticity and a casualty of the rejection of classicism, institutions relegated them to the basement, or simply dismantled and disposed of them.

Favour is swinging back, as it tends to, and curators are now pondering what was lost in slinging out the casts. Some were records of sculpture subsequently destroyed in conflict. The moulds preserve fragments of paint that came off during the casting process, long since 'cleaned' from the original. Casts made hundreds of years ago track the impact of environmental damage or over-enthusiastic 'restoration': they have become more authentic, in a way, than the depleted original.[xv] Having played a starring – and controversial – role in European art history, they have acquired an aura of their own.

#06

LINGBI

P U SONGLING'S 'THE ETHEREAL ROCK' IS AN UNCONVENTIONAL romance: a fable of obsessive love binding a rock collector and a mysterious stone he finds tangled in his fishing net.[i] The rock is entrancing: pitted, contorted, crannied and peaked. It has mysterious powers. Like the great mountains it resembles, clouds of steam puff around the rock's peaks ahead of rainfall. The marvellous rock becomes an object of envy and desire: a local despot orders his servant to steal it, but the stone escapes and tumbles into the river, eventually making its way back to the collector, who hides it away.

Rock collectors across the centuries have compiled rules according to which a fine specimen might be judged. The eleventh-century Chinese calligrapher Mi Fu suggested four categories: *shou*, fineness and elegance; *lou*, penetrability or routes for the eye into the rock; *tou*, holes and perforations; and *zhou*, wrinkles. Other criteria include a resemblance to painted mountain landscapes, romantic charm and a quality of fascinating ugliness.

For collectors, the most highly prized of dramatic rocks is the *gongshi* – known in English as a 'scholar's rock' or 'spirit stone' and in Japanese as *suiseki* – an object of evocative, dynamic form used as a focus for contemplation. Placed on carved wooden stands that hold them steady in their most captivating position, *gongshi* can be small as a thumb or tall as a teenager. Historically, the most valuable *gongshi* – and those which best meet the calligrapher Mi Fu's criteria – are *Lingbishi* (Lingbi stones): tortured, twisting fragments of fine-grained limestone from Lingbi county in the north of Anhui province. Lingbi stones are commonly black or dark grey, criss-crossed with thin seams of white.

Mi Fu's notorious petrophilia earned him the nickname 'Crazy Mi', but his obsession was understood to be honourable. On receiving a government post in Anhui province Mi took a detour from a journey to present his credentials to the local prefect, and instead donned official robes to visit a famous local rock, which he bowed before and addressed as 'elder brother'. The story became celebrated, and a popular subject for artists because of the philosophical position informing Mi's eccentricity: that the power and spirit of the natural world outweigh the ceremonies of public life.

Dramatic rocks have been integral to Chinese garden design since the Han dynasty (200 BCE–220 CE): they condensed the potent wildness of nature, and evoked the five sacred mountains that symbolise the centre of the Earth and its four corners.[ii] The tradition of scholars and artists naming special rocks dates back at least to the reclusive fourth-century poet Tao Qian. A particular garden rock he lay on when inebriated was bestowed the title 'sobriety rock'.[iii]

Lingbi and other prized scholar's rocks were mined from caves and muddy subaquatic environments. In truth, most *gongshi* have been subject to some titivation from sculptor's chisel and lacquer brush, but their forms are natural in origin.[iv] They are geological microcosms of the spectacular landscapes so important to Chinese art and literature: the huge deposits of limestone that transformed into caverns, subterranean waterways and towering pinnacles over centuries of erosion.

Lingbi limestone is so dense that it emits a bell-like tone when struck. The finest Lingbishi were white, and Hongwu, founding emperor of the Ming dynasty (1368–1644), had them distributed to temples as chimes.[v] Stone chimes from these cave complexes played an important role in the music associated with the earliest Chinese religious rituals. Texts from pre-Han and Han times use the term 'bell teats' for sonorous stalactites found in limestone caves. Later Confucian bells cast in metal retained their association with fertility and carried decorations symbolising the nipple of the female breast.[vi]

Stones started to migrate indoors during the Song dynasty (960–1279 CE), when evolving bureaucracy kept the learned elite tied to their desks and away from their country estates. For the literati – learned men who held government positions and excelled in poetry, painting and calligraphy – *gongshi* brought the spirit of mountain and garden into the scholar's studio. The finest examples offered a world within a world for contemplation. Some were used as incense stands, the smoke furling around the folds of rock like mist on mountaintops. Others were arranged in clusters, as tabletop landscapes. All allowed the mind to travel, providing a continuum between nature and the intellect considered fundamental to creativity.

Literati of the Song period nurtured a self-image as heirs to ancient cultural traditions. As the obsession for collecting and displaying rocks became fervent in the early twelfth century, Lingbi locals engaged in subtle rebranding to push the connection between their local limestone and the sacred chimes of old, replacing the original characters for Lingbishi – 'Solitary Cliff' – with the homonym 'Spirit Stones'.[vii]

Much as the wealthy Dutch of the 1630s developed a mania for tulips, so the literati of the Song period competed to pay outlandish prices for *gongshi* and stones for the garden. This wave of petromania reached its apex during the rule of Emperor Hui-Tsung (1100–26 CE) – painter, calligrapher, garden-maker and lover of stones. On Hui-Tsung's behalf, lakes were dredged, bridges dismantled and properties ransacked. The Emperor essentially bankrupted himself fulfilling his appetite for spirited rocks. By the time Pu Songling was writing, five centuries later, a second wave of petromania was commencing which would last well into the nineteenth century.

The Boston sculptor Richard Rosenblum became one of the great petromaniacs of the twentieth century, accumulating one of the world's largest hoards of *gongshi*. Writing in the 1990s, Rosenblum lamented the absence of a comprehensive collection of *gongshi* in China. Thanks to their associations with 'decadent' high culture, spirit stones had been targeted during the Cultural Revolution. Many had been concealed in gardens and returned to a less venerated identity, simply as rocks.

He also thought the stones unfairly neglected in the West, and collectors blinded to their charms by centuries of prejudice: in the European tradition creativity is considered the preserve of man, who was superior to, and lord over, nature. Rosenblum intended to address this. On his death in 2000, some three dozen examples from his prized collection went to the Metropolitan Museum of Art in New York, including a captivating, mountainous specimen in black Lingbi currently on display in gallery 215. For want of a more apt category in the museum's archiving system, it is catalogued as 'sculpture'. Western traditions of connoisseurship are still challenged by the idea that we might appreciate the work of nature as art, to the point of love or obsession.

In Pu Songling's fable, the collector is visited by a stranger, an old man who demands the return of 'his' rock. To prove his claim, he tells the collector that the largest of the stone's ninety-two crannies carries a tiny engraved inscription: 'Offered In Worship, Ethereal, The Celestial Rock'. The two men begin to bargain, and the collector offers three years of his life in return for the rock. The stranger pinches shut three of the rock's crannies, leaving eighty-nine: the age at which the collector will now die. He is buried with his stony paramour.

#07

ONYX

'ONYX, WITH ITS DENSE DEEP BLACK, IS A SPECIAL SUBSTANCE,' decides Roger Caillois in *The Writing of Stones.*[i] Published in 1970, late in the French philosopher's life, this idiosyncratic tome is a meditation on his collection of mineralogical eccentricities, stones he knew intimately, both by sight and by touch.

As a young man, Caillois had been a surrealist, and a great friend of André Breton, who was both the father of Surrealism and much prone to falling out with people. The inevitable happened in 1934, and Caillois and Breton parted ways after disagreeing over Mexican jumping beans. Observing the beans' jerky, erratic movements, Caillois was convinced their motion was caused by larvae and wished to dissect them: Breton upbraided him for failing to enjoy the poetry of a miracle without wishing to pick it apart.[ii]

Caillois was interested in the innate beauty and 'magic' of natural phenomena. Unlike the surrealists who looked within, to the subconscious, his curiosity led him to ponder the ways and structures of the world. Here, too, there was space for wonder and miracles. In *The Writing of Stones* he invites his readers to step away from the nineteenth-century drive to categorise, and the geological urge to analyse, and instead indulge in pareidolia – the innate human tendency to detect image and pattern in random marks.

Much of Caillois's mineral collection was dedicated to cryptocrystalline varieties of quartz: agates, chalcedonies, jaspers and banded onyx.[iii] He teases his imagination with the stones, testing whether his memory-saturated mind will take the bait and search out an image. In some he finds a landscape, in others a misty forest. Occasionally a face emerges. The drama of black and white onyx, sliced and polished, seems to him transcendent, creating a graphic language all of its own.[iv]

Caillois's onyx specimens included a sliced black oval enfolding a white half circle, golden at its upper margins, like a view of after-noon sky from the mouth of a cave. Another seems to hold, frozen in time, a chick breaking from its egg. 'A pair of thin lips, insipid yellowish white against an intensely dark background' suggest, pale, lithic *labia majora*.[v] In the most dramatic, firm white lines stand out within the oily blackness of the stone as though calligraphic script thrown out from the bowels of the Earth.[vi]

Novelist Marguerite Yourcenar, writing of Caillois's calligraphic onyx, points out that those traces 'which sometimes almost exactly resemble writing . . . actually do transcribe events from millions of years ago.' Just as Caillois finds in his picture stones an art without artists, so these 'authorless inscriptions may be regarded as a first draft of a chronicle of stones'.

This sense of stones' lively qualities was shared by Roman lapidaries centuries before who wrote of a pregnant Earth, and her veins, arteries and entrails. Stones also coupled and reacted to one another. Sparks flew when male and female flints met – the Romans called them *lapis vivus*, living stone.[vii] Onyx apparently mated with sard to produce banded brown and white sardonyx, the finest hardstone for carving cameos.[viii]

Hardstones – semi-precious stones including agate, onyx, sard and carnelian – have been carved since the seals of ancient Mesopotamia. Until the fifth century BCE, they were worked *intaglio* – engraved, with the image cut into the surface. This was fine for a seal, since a clear, raised version of the engraved form appeared, as desired, when pressed into clay or wax, but for a jewel, it offered an indistinct image. The Greeks were the first to sculpt stones in projected relief – *anaglyphs* – the antecedents of the cameos so popular in first-century Rome. Cameos, like intaglios, are worked with the side of a drill dipped in abrasive powder.

Cameos in onyx and sardonyx use the translucent stones' banded colour to dramatic effect. A portrait (c. 41–54 CE) of Augustus shows the Emperor in profile with his muscular back turned towards us, sword raised, draped in a ceremonial cape. His radiant white body – suggesting the moral probity associated with *candidus*[ix] – stands out against the tobacco-coloured ground of the stone, so finely carved that we can make out decorative details of his cape. The ribbons securing his chaplet of oak leaves flutter, meltingly pale, carved into the faintest trace of white banding above the sard.[x]

Some use multiple bands of colour. A sardonyx cameo of Aurora, goddess of dawn, driving her chariot picks out the nearest horse in an upper layer of golden brown, as though emerging from darkness into the coming light of day.[xi] It was a popular design and copied, less skilfully, by other carvers: cameos had amuletic powers, and honouring Aurora brought luck for the day ahead.

Cameos' periodic return to fashion is close coupled with associations of Imperial Rome: both the Holy Roman Emperor, Frederick II, in the thirteenth century, and Napoleon in the early nineteenth, encouraged the glyptic arts. For Frederick, as for the Romans, cameos were talismanic: a carving of Hercules wrestling the Nemean Lion was prized for its power to bring victory in battle. Napoleon had a number of cameo portraits carved – complete with a chaplet of laurel – in clear echo of Augustus.

The taste for cameos spread to his sisters and across the ladies of Napoleon's court, who wore tiaras and necklaces set with carved onyx and sardonyx jewels.[xii] What was à la mode in Paris set fashions across the rest of Europe: there was a craze for cameo jewellery in the early nineteenth century, furnished by expert carvers in Rome and at Idar-Oberstein in Germany.

Not for Caillois, all this power and preciousness. What interested him were not jewels of quantifiable value, but stones so curious they suggested an 'alien reality'. In banded onyx and its siblings, he saw 'a pre-existing general beauty vaster than that perceived by human intuition' which he prized all the more for being the result of impurities, inclusions, erosion or serendipitous breakage. Unlike wrought cameos that transform onyx and sardonyx into sacred talismans or glorifying portraits, the natural formations in his collection 'owe nothing to patience, industry, or merit,' he wrote. 'Each stone, as unique and irreplaceable as a work of genius, is a valuable at once pointless and priceless, with which the laws of economics have nothing to do.'[xiii]

#08

PINK
ANCASTER

B ARBARA HEPWORTH HAD NO PATIENCE WITH THOSE WHO marvelled that this bird-like woman could drive cool elegance from brute rock. 'They still think of sculpture as a male occupation: because, I suppose, they have a misconception of what sculpture involves,' she told critic Robert Hughes, exasperated, after he stumbled into 'little woman' pleasantries on a visit to her studio in St Ives, Cornwall, in 1966. 'There is this cliché, you see: a sculptor is a muscular brute bashing at an inert lump of stone, but sculpture is not rape. No good form is hacked. Stone never surrenders to force.'[i]

Decades earlier, Hepworth and her friend Henry Moore had helped redefine the relationship between sculptor and material: they favoured 'direct carving' rather than working first in plaster or clay. Sculpture, in stone or hardwood, was a collaborative process between artist and material. 'Each material demands a particular treatment,' Hepworth wrote in 1930, 'and there are an infinite number of subjects in life each to be re-created in a particular material. In fact, it would be possible to carve the same subject in a different stone each time, throughout life, without a repetition of form.'[ii]

For Hepworth, stone was not an inert substance: she responded powerfully to the material itself, its particularities and associations. Growing up in Yorkshire, she had accompanied her father, a structural engineer, all over the West Riding.[iii] She felt a familial bond to the landscape and the rocky forms exposed upon it. In her sixties, shortly before Hughes's visit, Hepworth commissioned a suite of photographs of Yorkshire – fields boundaried by dry stone walls, the Cow and Calf rocks above Ilkley – allowing her mind to travel back to the landscapes of her childhood (see: Millstone Grit, p. 250). She responded to rock formations with the same ardour she might afford a marble head glimpsed in Rome.

In September 1931, when she was a young mother, awkwardly married, she invited the painter Ben Nicholson to join artist friends on holiday in Happisburgh in Norfolk. Earlier that year, Hepworth's sculptures had been shown coupled with Nicholson's paintings: she felt at the time 'something momentous has happened'.[iv]

The invitation to Norfolk, then, was a gesture of seduction, one she baited with photographs of rocks: 'They have moved me more than anything I have ever seen sculpturally. We took a boat out onto the western islands and there were purple rocks, groups of rocks, hundreds of rocks all worn by the water and yet retaining their fundamental thrust and we sailed round watching the changing aspects and the revealing of surprise movement.'[v]

The fundamental thrust worked its magic: Nicholson came to Happisburgh, and left enraptured. Afterwards, Hepworth wrote to him in lithic passion: 'your dear head is like the most lovely pebble ever seen and your thoughts clear as the pebbles just left by the sea and I love all you are and do.'[vi]

Hepworth and Nicholson's love may have felt clear as a pebble, but things were less smooth for others involved. Hepworth requested a divorce, provoking unkind reports in the newspapers. Nicholson attempted to balance this new relationship with active fatherhood to his three young children with the painter Winifred Nicholson.

A modern ménage emerged, with Nicholson dividing time between wife and lover – a situation tolerated partly because it accorded with their progressive ideas on art and life, and partly because all three were Christian Scientists. The religion's tenets forbade adherents from harbouring ill feelings towards those they loved: control 'evil thoughts' lest they control you.[vii] A cynic might venture that the prohibition on rancour played greatly to Nicholson's advantage.

Inspired by visits to Picasso, Mondrian, Arp and others in Paris, both Hepworth and Nicholson started to move towards abstraction in their work. A transformative moment came in 1932 when Hepworth pierced a hole through a small sculpture in pink alabaster[viii] – describing the experience as one of intense pleasure.

Between 1933 and 1934, Hepworth made a series on the theme of mother and child: organic, globular shapes like small eroded boulders. Many would be destroyed in the coming war. Among those that survive is perhaps Hepworth's most touching work, *Mother and Child* (1934), carved in pink Ancaster, a compact oolitic limestone from Lincolnshire. It is small, about thirty centimetres high, the stone speckled and a dilute, membraneous pink: the colour of connective tissue, tender, pale parts of the body's inner workings.

Pink it may be, but it lacks the gelatinous fleshiness of alabaster or chalcedony. Ancaster was predominantly a building material: a practical British stone used in the same period by Moore. Art historian Anne Middleton Wagner suggests that Moore and Hepworth selected the stone not because it suggested flesh, but precisely because it did not. No silky marble skin here: the unmistakable rock surface of the Ancaster limestone spoke in broad tones of the hills from which it was quarried.[ix]

Mother and Child is in two parts: the 'mother' is rounded like an old mountain, with a valley forming her lap. In this stands the small figure of the 'child', independent, yet close, leaning in toward her. Hepworth offered a new interpretation of an ancient subject – one for an era of progressive relationships and Freudian psychoanalysis – two individuals, cut from one stone, but ultimately separate. It was a radical, if idealistic statement, and one perhaps provoked by Hepworth's discovery that she was pregnant by Nicholson.

Hepworth was not independently wealthy, but her work already attracted collectors. She was able to support herself – just – as a divorced mother, during the lean 1930s. Pregnancy left her feeling drained. She could no longer work on heavy sculpture with the baby so active she felt it was 'playing a tennis match inside me'.[x]

Little wonder: it was not one baby, but three. All delivered, tiny but alive, on 4 October 1934. Nicholson, still dividing himself between Hepworth and his family with Winifred, left for Paris a few months after the birth. Still very weak, Hepworth was terrified – how could she look after four children and work to support them? The situation was unsustainable. The gas leaked, the boiler leaked and there wasn't enough space. Letters reveal her torn between intense love for her children, bone-tiredness, hunger to make art again and panic that she might not.[xi] Her solution was to place the triplets for a few months in a nearby nursery training college – Wellgarth – so that she could start working again while they received round the clock care.

One of Hepworth's first sculptures after the birth was *Three Forms* (1935): two ovoid shapes and a sphere on a thick base, all in white marble. The idealistic earthiness of *Mother and Child* had transformed into a geometric study in tension and relationships. If the forms are the triplets, each is distinct and apart, and she as the mother, the supportive base holding everything together.

#09

QUARTZ

THE CATALOGUE ENTRY FOR THE LIFE-SIZED ROCK CRYSTAL skull at the British Museum demurely notes that its origins 'are most uncertain'. Purchased in 1897, from Tiffany and Co. in New York, the skull is spring-water clear, carved from a single huge quartz crystal, and was believed Aztec in origin. Gemologist George F. Kunz described its provenance in *Gems and Precious Stones of North America*: the skull was apparently 'brought from Mexico by a Spanish officer sometime before the French occupation of Mexico',[i] and had thereafter passed through the hands of a number of European collectors.

Kunz, conveniently, was also vice-president of Tiffany's. In a note to Sir Charles Hercules Read, keeper at the British Museum, Kunz confessed he would be 'much pleased to have your institution possess this remarkable object'. The honey worked its magic. Why should the museum not buy a crystal skull? These haunting objects were on the shopping list for late nineteenth-century collections. There is one at the Musée de l'Homme and three at the Musée du quai Branly in Paris. Over a century later, another – purportedly from the collection of nineteenth-century Mexican president Porfirio Díaz – was donated to the Smithsonian Institution in Washington DC.[ii]

Quartz – silica – is among the most common minerals in Earth's crust. Most sand is quartz. Agate, flint and chalcedony are cryptocrystalline varieties, composed of microscopic crystals. Quartz is present in granite and, of course, sandstone. Since prehistory, rock crystal – colourless macrocrystalline quartz – has been associated with magic and power, treasured for ritual use from Australia to South Africa to the Americas. Its numinous qualities include water-like transparency (which gave rise to an enduring belief it was preserved ice) and the production of flashes of light under friction or stress (triboluminescence). Fire can be produced from both quartz sparks and quartz lenses. 'This stone, set for a short while in the sun, makes fire and light that sets dry toadstools alight,' according to the fifteenth-century *Peterborough Lapidary*.

In 1992, curator Jane Walsh took delivery of the skull at the Smithsonian. Unconvinced by its provenance, she started poking into the backstory of its rock crystal siblings in other collections. A peculiar pattern emerged. All had appeared within a thirty-year period, between 1860 and 1890, and all at some point had passed through the hands of one man: Eugène Boban. In 2008 – the very year the unfortunate fourth film in the Indiana Jones franchise 'revealed' that crystal skulls had been planted on earth by aliens – Walsh and colleagues at the British Museum conducted tests on the skulls in both collections, comparing them with a verified pre-Columbian quartz goblet.

Rock crystal is difficult to work. It's hard (7 on the Mohs scale) but you can't chisel it like a lump of rock because it will crack. It must be ground with abrasive powder from a harder substance: corundum (the stuff of rubies and sapphires) or diamond. This is painstaking, highly skilled work. Very few pre-modern cultures could form and engrave rock crystal. The huge chunks used in church regalia in medieval Europe were simply polished and left lumpen. Theophilus Presbyter, writing in Cologne in the early twelfth century, believed in order to carve rock crystal 'it had first to be heated in the cardiac cavity of a newly slaughtered goat.'[iii]

The great masters of rock crystal lay to the south of the Mediterranean, in Fatimid North Africa, and in particular Carthage, in present day Tunisia.

Walsh and her colleagues concluded that neither crystal cranium was pre-Columbian. The fine work and high polish indicated modern lapidary tools. The one belonging to the Smithsonian was probably made in the 1960s, shortly before the donor acquired it in Mexico – it had never been owned by Porfirio Díaz. Inclusions in the quartz of the British Museum skull pointed to a crystal source in Brazil, Madagascar or possibly the European Alps.

Walsh found a letter in the Smithsonian archive detailing Eugène Boban's attempt to sell a crystal skull to the Museo Nacional de Mexico in 1885, assisted by Leopoldo Batres, the Porfirio Díaz-appointed inspector of archaeological monuments. The Mexican museum rejected the skull, suggesting it was modern European glasswork. Batres then denounced Boban, who abruptly departed (with the skull) for New York. Walsh's investigation of the mysterious Frenchman took her down a mucky trail, exposing a lawless era for both archaeology and museum acquisitions.

Boban was an expert scholar of Mesoamerican artefacts and codices, who lived for long periods in Mexico City in the second half of the nineteenth century. During the catastrophic French intervention (1863–7), Boban anointed himself antiquarian to the Emperor Maximilian (on the basis of a single visit the Emperor paid to his shop). In 1865, he ascended to the French scientific commission, and amassed a collection of thousands of pre-Columbian artefacts, which were exhibited in Paris in 1867.[iv]

Boban was a trusted source of Mexican antiquities: he supplied and advised museums in Europe and North America. In 1875 he sold his own collection to Alphonse Pinart, the French linguist and ethnologist. Pinart in turn donated it – crystal skulls and all – to the Musée d'Ethnographie du Trocadero, now the Musée de l'Homme, apparently in return for government sponsorship of his overseas expeditions.[v]

In 1900, three years after the British Museum paid a handsome £220 for their skull, Boban sounded a note of caution in an interview: 'Numbers of so-called rock crystal, pre-Columbian skulls have been so adroitly made as almost to defy detection, and have been palmed off as genuine upon the experts of some of the principal museums of Europe.'[vi] The statement was disingenuous in the extreme. The crystal skull, as a category of ancient Mexican artefact, was one entirely invented by Boban himself. The Aztecs did not carve crystal skulls, ergo the examples he sold were not copies or fakes – they were fantasy.[vii]

Boban's bogus skulls were carved from real rock crystals. In the twentieth century, synthetic quartz made real magic, bestowing a dazzling corona on deities – living and dead – worshipped by millions in the modern era.

In 1913, a kid from Kyiv called Nuta Kotlyarenko moved to the US and had his name mangled by emigration. By the 1930s, he was making rhinestone-encrusted G-strings and nipple pasties for New York's burlesque performers. The following decade, known now as Nudie Cohn, he moved west and took his rhinestones with him, setting himself up in Hollywood as Nudie's Rodeo Taylors.

Rhinestones were originally just as their name suggests: little quartz pebbles from the River Rhine. As long as glass has been available, jewellers have fabricated gemstones, but faceted synthetic crystals were a phenomenon of the late nineteenth century with Daniel Swarovski a driving force. Rhinestones dazzled amid the electric lights and cut glass of the jazz age: on stage, they were mesmerising.

Nudie's Rodeo Taylors peddled a dazzling peacock style for men that by the 1970s made the term 'rhinestone cowboy' synonymous with music stardom. In 1956, Colonel Parker commissioned Nudie to make a gold lamé suit with rhinestone-encrusted cuffs and lapels – valued at a headline-grabbing $10,000 – for his breakout star Elvis Presley.

Nudie went on to rhinestone everyone from Dolly Parton to Keith Richards to The Flying Burrito Brothers. In 1969 *Rolling Stone* portrayed fatal bad boy Gram Parsons in his custom-made white Nudie suit: naked women on the lapels, front and sides splashed with rhinestone outlines of the hemp leaves, opium poppies and barbiturates that would carry him off four years later. The reporter marvelled at how such camp, crystal-decked flamboyance had become synonymous with ultramasculine swagger: 'Nudie somehow has managed to convince nearly 25 years of rough and rugged cowboy types they should buy blue boots studded with costume jewelry and suits of magenta elastique dripping with rhinestoned fringe.'[viii] A crystal skull would have fitted right in.

#10

RED
OCHRE

IN LIMESTONE CAVES ACROSS THE WORLD, A GLOBAL MENAGERIE IS painted in red and black. A 45,000-year-old warty pig, with his double chin and finely rendered bristles, in Sulawesi, Indonesia; a portly, 28,000-year-old kangaroo at Kakadu in Australia's Northern Territory; thundering herds of bulls and horses at Lascaux in France; a yet undated (but older than was imagined possible) armadillo, capybara, jaguar and tapir in Serra da Capivara in north-east Brazil; a magnificent bison at Altamira in Spain; a fighting lion at Bhimbetka in India: all carry the intense red of hematite.

The earliest known paint – long predating these cave paintings – was red as blood. Ochre, bone and charcoal, mixed to a paste in the bowl of an abalone shell, it was found in the Blombos Cave on the coast of South Africa's Southern Cape. The cave was in use perhaps 100,000 years ago, and the paint was made with care and knowledge: raw ochre, grindstones, palettes and shells found on the site suggest routine manufacture.[i] The pigment had been heated and pounded, made into a sticky, penetrating paste, perhaps with bone marrow and urine.[ii] What, or more aptly, who, was being painted is not known. The cave has also yielded the earliest known piece of 'art' – ochre again, but this time a carved fragment of stone, small enough to cup in the hand, incised with crossing lines.

Red ochre is a broad term for earth pigment, usually from hematite – 'blood stone' – an iron oxide, found as a rusty mass or in metallic lumps that leave a sanguine streak when rubbed on a touchstone. Iron oxide painted the landscape long before it was used on bodies, hides or walls: hematite streaking the rocks and waterways would have shone out to ancient eyes. As modern humans spread across the world from Africa 70,000 years ago, they carried their affinity for red ochre with them through Asia, Oceania and eventually Europe. It's an attraction that can be traced back to our ape ancestors and the trichromatic vision we share with them, allowing us to see red standing out against a green background.

Tens of thousands of years before our ancestors braved the subterranean darkness to commit visions of animal life and spiritual transformation to rock surfaces, ochre was processed and used in other ways. The earliest evidence for human engagement with hematite has been found at Olorgesailie in Kenya, in fragments worked over 300,000 years ago.[iii] We are not the only large-brained hominids to be smitten. In the Netherlands, fragments of ochre have been found among Neanderthal remains that predate the arrival of our African predecessors in Europe.

The ability to find, clean, heat, grind, sift and blend ochre with a binding medium tells us that our ancient ancestors were able to plan, and to process material. They evaluated the qualities of different raw ingredients, and prized those from some sites over others.[iv] Shells stained with ochre, perhaps used as body ornaments, or to hold pigment, were found at a 92,000-year-old burial site at Qafzeh Cave in Galilee.[v] Alongside stone tools, the digging for and processing of ochre is arguably the earliest exploitation of earth minerals, and the use of heat to modify its colour, the first technological use of fire.[vi] Hematite stained the Paleolithic red.

Ochre paste played many roles. It could be used as an adhesive, and to tan hides. Applied in paste to the body it protected the skin from overexposure to the sun, and helped repel biting insects, allowing our ancestors to travel further and forage more widely. It has even been suggested that pregnant women consumed ochre, fortifying the baby in utero with iron.[vii] Ochre burials of the Upper Paleolithic, such as the Red 'Lady' at Paviland in South Wales (famously misgendered by William Buckland),[viii] may be the result of red paste applied to garments which have long since disintegrated, leaving nothing but pigment to colour the bones.

It is misleading to make a clear distinction between 'functional' and 'ritual' applications of ochre, just as it is to understand cave paintings as 'art' in the modern sense. Ochre paste applied to the skin could at once be decorative (the human eye is drawn to red), practical (sunscreen), and have important spiritual or ritual connotations. It rendered the body very visible, perhaps broadcasting amity: here were no hostile intentions requiring concealment.

The processing of this valued material is likely to have had tremendous ritual significance. Paint was not merely 'paint', writes David Lewis-Williams of the Rock Art Research Institute in Johannesburg. 'It had supernatural properties and significances.'[ix]

Twenty years ago, Lewis-Williams presented a radical theory

about the origins of art. Humans in the Upper Paleolithic were not painting a record of the world around them on the cave wall: these first artists were shaman-like figures, and the images of animals and spirits were highly codified, the results of visions and induced hallucinations.

We modern humans are equipped with the same brain and nervous system as these ancient ancestors, so Lewis-Williams looked to recent history for clues as to how the people who painted the caves at Altamira and Lascaux might have understood their relationship to the material and spiritual realms. This was not uncontroversial: archaeologists objected to Lewis-Williams looking to the modern shamanism of Siberia, southern Africa and South America to throw light on Upper Paleolithic rituals and beliefs.[x]

Nevertheless, his theories about the origins of art, and the roles played by both the ochre and the cave surface, are thrilling. Lewis-Williams reads the cave as a transition point between the physical and the spiritual realm, accessible only to powerful figures who could act as intermediaries. Intense under any circumstances, lit by flames and animated by musical rhythm, these cavern walls would have seemed to vibrate: perhaps they were places of hallucination rituals or vision quests. The painted images have a tight relationship to the rock surface: the animal figures follow its smallest contours, suggesting the artist selected their sites in part by touch. As a result, the images appear to emerge from or float on the rock, as though they were an intrinsic part of the cave that has simply been revealed by the application of ochre and charcoal paste.[xi]

The image of the human hand is near ubiquitous among painted caves of the Upper Paleolithic. Often, they appear in silhouettes, made by blowing a mist of red ochre over a hand held against the cave wall. Rather than simple marks left by primitive minds, Lewis-Williams reads these as the traces of an intense act of binding, blending hand and wall into a single surface, repeated over and over as part of the deep ritual of the cave. The blown ochre – this powerful, supernatural substance, red as birth – seals the human body to the spirit realm beyond.

STONE
TECHNOLOGY

COADE STONE

COAL

COLTAN

FLINT

HAÜYNE

LODESTONE

MILLSTONE GRIT

MICA

OBSIDIAN

STONE
TECHNOLOGY

IN PRIMO LEVI'S AUTOBIOGRAPHICAL STORY 'IRON', THE YOUNG chemist learns to understand matter by conquest: first with the guidance of his professors amid the flames and flasks of his university laboratory; later with a fearless friend, clambering up the rocks and ice of the surrounding mountains. Chemical processes, for Levi, become a metaphor for other transformative encounters. 'It might be that Matter is our teacher,' he writes. 'And perhaps also, for lack of something better, our political school.'

Urstoff – the 'primary matter' of Levi's wild rockface – is a harsh professor. It is not a tamed element in the analytical lab that will respond, more or less compliant, to routine experiments. Instead, the Alpine rock teaches Levi about the limits of human ability, and the body's capacity for endurance: necessary skills to survive in an environment over which they have little or no control.[i]

Humans' first attempts at mastery came through stone. It provided our earliest technology. Hammer stones came from any lump of hard stuff you could get a good grip on. About two and a half million years ago, in the territory of present-day Ethiopia, hammer stones were used to flake off fragments from other stones, forming the earliest sharp-tipped chopping tools. Later came hand axes, the first blades and scrapers, and then slivers of chert and flint that were fitted into wooden handles as harpoons or spears. Over hundreds of thousands of years, these tools became specialised, refined and beautiful, treasured as sacred as well as functional objects. Rock provided building materials for early settled societies.

Querns from rough sandstone were used to pulverise grain and make it edible.

Technological developments occurred at different times in different places, but what we broadly refer to as the Stone Age extends over about 3.4 million years. The Paleolithic started around the time the genus *homo* emerged, and the end of the Neolithic came on gradually, between 6,000 and 4,000 years ago with the development of the first bronze tools. Most of human history, in other words, took place in the Stone Age. Knowledge of metalworking did not drive out stone tools: instead it expanded the range of available materials, and the means to work them. Well into the modern era, human culture used stones to grind flour, sharpen metal blades and spark fires.

We still live in a stone age. Fifty billion tons of sand and gravel are mined for use in construction every year, a quantity that dwarfs all other extractive industries (as a point of comparison, the amount of useable iron ore mined is about 5 per cent of that: 2.5 billion tons.[ii]) The most elegant and otherworldly expressions of our contemporary technological age – obsidian-sleek cameraphones and flat-screen monitors; computer hard drives; wind turbine generators; electric car batteries – all rely on 'rare earth' elements[iii] mined and processed from a handful of uncommon igneous rock types.[iv]

#01

COADE
STONE

A T THE TOP OF LYME REGIS, BELMONT STANDS SQUARE AND pretty as a cake napped in pink fondant, iced with a delicate hand. On the façade of this maritime villa, heads of Poseidon and Salacia gaze in cool splendour from arches set with squirming dolphins between blocks shaped like rough-hewn stone. There are friezes of swagged oak leaf garlands, and urns stud the parapet like birthday candles. Standing high above the bay, Belmont looks out through improbable palm trees down to the curving stone harbour wall of the Cobb, site of Louisa Musgrove's fall[i] and Sarah Woodruff's lonely vigil.[ii] Belmont was home to the novelist John Fowles, who took Lyme as the setting for *The French Lieutenant's Woman*, but its façade was confected by a much earlier resident, and this display of crisp, durable architectural sculpture performed as an advertisement for her work.

Eleanor Coade was born in Lyme Regis in 1733, but the business that carried her name was based in Lambeth, on the south bank of the Thames, opposite the Palace of Westminster. Still marshy and rural, Lambeth was a manufacturing district where terracotta, glass, stoneware and soft paste porcelain were produced at easy distance from ships that deposited raw materials and collected finished stock. In 1769, following the death of her bankrupt father, Eleanor and her mother (another Eleanor) went into business with Mr Pincot, a failing manufacturer of artificial stone. Within two years, Pincot had been ousted, and the artificial stone rebranded as Coade's Lithodipyra – a grand Greekish coinage with a nod to the product's double firing. It would later be known simply as Coade Stone.

While she likely refined Pincot's recipe, Eleanor Coade did not invent the product that carried her name. What she brought to the business was promotional flair and an eye for architectural fashion. The Royal Academy of Arts had been founded in 1768, and she raised her product's prestige by association with prominent sculptors, among them Thomas Banks, John Flaxman and John Bacon, who was awarded a gold medal and made associate academician in

1770. Coade exhibited the best work of her superintendent sculptors at the Royal Academy. To architects she offered an illustrated catalogue of 788 designs – faces, figures and urns, friezes, columns, capitals and chimney pieces – all adaptable, and made to order.[iii] Used by the great architects of the day, the stuff became ubiquitous on Georgian buildings of a certain ambition – in London that included Buckingham Palace, Sir John Soane's house at Lincoln's Inn Fields and the Old Royal Naval Palace at Greenwich.

Coade Stone 'may not have revolutionised the architecture of the late Georgian period in any structural sense, but its aesthetic contribution was much more considerable than is generally realised,' wrote Roger White, secretary of the Georgian Group in 1990. This reliable, elegant and affordable product allowed the great architects of the day to indulge in an orgy of classical ornamentation that would have been prohibitive in carved stone.[iv]

Described as an artificial stone (though one might more accurately call it human-assisted), Coade was a mixture of pale West Country clay, crushed flint, fine sand and waste products conveniently generated by neighbouring manufacturers: finely ground fired stoneware and soda glass. The formula was guarded, but not patented: there was probably no single master recipe, and proportions were adapted to fit the scale and detailing of each piece, whether a staunch lion or a copy of the Borghese Vase. Bacon or one of the other sculptors commissioned by Coade first modelled a form in clay – a little large, to allow for shrinkage – from which a plaster mould was taken. This mould could be re-used and adapted – a sheaf of wheat added here, a dolphin there. Sheets of raw Coade stone were pressed into the interior of the mould and then fired for four days in a coal-fired kiln at 1,100 to 1,150 °C.[v] In proclaiming her product superior to natural stone, Eleanor Coade did not overstate its durability: after 250 years, the sculptures and exterior mouldings produced under her name have remained bright and finely detailed, less prone to erosion and pollution damage than the Portland Stone they emulate.

Made precisely to order, Coade stone sculptures were exported and examples survive in Russia, South Africa, Canada and Brazil.[vi] The product's success was not only the result of Coade's business nous: it was also a brilliant coincidence of timing. Artificial stones have been produced for millennia (see also Gypsum, p. 187): the earliest concretes found in Greece date to 1300 BCE, and lime plasters go back even further, to the Neolithic settlements at 'Ain Ghazal in present-day Jordan. Both concrete and plaster are air-dried, rather than fired, and neither offers the combination of durability, monumentality and ability to show fine detail of Coade stone. In London, other manufacturers had experimented with similar products with limited commercial success. Coade stone's launch coincided with the neoclassical designs of Robert Adam and John Soane, and a fresh appetite for caryatids, friezes and urns. 'The feeling for the externals of architecture changed, after the arrival of Adam, from the grimness of a mask to the delicacy of a feminine "make-up",' wrote John Summerson, curator of Sir John Soane's Museum. 'Stucco and Coade stone have a slightly cosmetic character; they suggest, faintly and agreeably, the artificiality of powder and rouge.'[vii]

The manufactory did not survive long after the death of its proprietor: Eleanor Coade died in 1821, and the company that carried her name had crumbled by the end of the 1840s. The villain was probably changing tastes: Coade stone was associated with elite architecture of the previous century. By the time of the coronation of the eighteen-year-old Queen Victoria, in 1838, it felt outmoded.

It took patience borne of great respect to reformulate Eleanor Coade's recipe almost 200 years after her death. Materials are transformed by their passage through the kiln, so no simple analysis could generate the recipe in retrospect: instead stone carver Philip Thomason relied on trial and error to refine a formula over thirty-five years.[viii] Living near Lyme Regis, Thomason experimented with materials shipped from the region to London in the eighteenth century. His recipe uses whitish Devon ball clay, with local beach sand and flint, ground glass and stoneware. The logistics of firing and manipulating objects of this weight and size is, he reports, formidable. Much of Eleanor Coade's stock would have been damaged and discarded. Today, Thomason's formula is a close enough match to restore Coade stone sculptures damaged by the deterioration of metal fixings, including those on the façade of Belmont.

#02

COAL

ETH STEPHENS'S FAMILY HAS BEEN IN MINING FOR GENERATIONS. In the seventeenth century her forebears dug tin in Cornwall. Later they crossed the Atlantic, and brought old-world expertise to mining copper in Michigan. In the 1940s they travelled to West Virginia, and the coal deposits of the Appalachian Mountains.

The migration routes of working populations have, for centuries, been drawn in grim, glitter-black lines of coal dust. Coal mining, and the smelting and manufacture it fuelled, drew rural populations to expanding cities. It powered the engines to move coal from pit to factory. The first steam-powered public railway line – the Stockton and Darlington Railway, opened in the north of England in 1825 – mainly ran freight, mostly coal. In the US, the Union Pacific Railroad was routed close to coal deposits – both fuel and cargo.[i] An earlier steam engine developed by Thomas Newcomen powered pumps rather than locomotives. These, too, were first used in coalmines, pumping water out of the deepening pits.

Stephens didn't go into mining, and instead moved to California, becoming a filmmaker, performance artist and educator. In 2008, Stephens and her wife Annie Sprinkle[ii] married the Earth in a public ceremony surrounded by giant redwood trees. 'Green Wedding to the Earth' was part of a seven-year cycle of ritual marriages. What started as resistance to homophobic legislation quickly pivoted into a gloriously unconventional form of environmental protest. In front of an audience dressed in green, Stephens and Sprinkle declared themselves 'ecosexuals' and proposed an alternative relationship to the Earth. 'People often think of the Earth as Mother Earth. But these days the Earth is so battered, abused, exploited, polluted, blown up and ripped apart that she can't handle the burden of being a full-time mother anymore,' they declared. 'Perhaps it would be better to imagine the Earth as a lover, because we tend to take care of our lovers instead of expecting them to take care of us.'[iii]

Stephens and Sprinkle's declaration echoes horrified accounts of Britain's industrial heartland 150 years ago. 'The earth seems to have been turned inside out. Its entrails are strewn about,' wrote scientist James Nasmyth after passing through the Black Country.[iv] 'Nearly the entire surface of the ground is covered with cinder heaps and mounds of scoriae. The coal, which has been drawn from the ground, is blazing on the surface ... Amidst these flaming, smoky, clanging works, I beheld the remains of what had once been happy farmhouses, now ruined and deserted. The ground underneath them had sunk by the working out of the coal ... Vulcan has driven out Ceres.'[v]

Mining, burning, and processing coal was toxic to the surrounding environment, and the exploitation of the Earth extended to the people labouring within it. An 1842 British parliamentary report on children in coalmines records some starting work as young as four, but eight to nine being the 'ordinary age' employment commenced. It includes an interview with a girl of unspecified age who laboured down the mine pushing laden coal waggons. The girl's account of long hours, physical abuse, meagre rations and relieving herself in the pit under the gaze of male workers is supplemented by a description of her shivering from cold: 'The rag that hangs about her waist was once called a shift, which is as black as the coal she thrusts, and saturated with water.' The inspector notes sudden embarrassment in the public house during this interview, mortification 'that these deeds of darkness should be brought to light.'[vi]

In modern cities, coal mining, its danger and devastation, can seem remote in time and space. Electricity appears – in a piece of conjuring so quotidian we no longer pause to marvel – at the flick of a switch. Coal is very probably in the mix. In the UK, about 3 per cent of electricity comes from coal-fired power stations, and 25 per cent from renewable sources.[vii] At the time of writing, China consumes over half of the world's coal-fired power, which supplies 57 per cent of its total energy use.[viii] Thirty per cent of all electricity consumed in the US comes from coal-fired power stations, though the picture varies state to state. Maine gets three-quarters of its power from hydroelectric, biomass and wind; in West Virginia 93 per cent of the electricity comes from coal.[ix] Vulcan is still driving out Ceres, he's just doing it in that magic place called 'away' – out of sight and out of mind for the world's affluent populations.

Stephens proudly describes herself as a hillbilly. Soon after she married the Earth, she travelled back to West Virginia, the landscape that formed her. Stephens thought the Appalachians would be there forever, but, as she explains in her documentary *Goodbye Gauley Mountain* (2013), a violent method of coal extraction called Mountain Top Removal 'is forever altering the landscape I call home.'

Coal was discovered in West Virginia in 1742 and extracted from underground mines: tough, claustrophobic work, carrying terrible risks to life and health, but a source of both employment and considerable pride. Mountain Top Removal is a recent – more profitable – practice. The name offers a bald description: explosives are drilled into the mountain, which is blown to rubble in increments until a coal deposit is exposed. Trees, undergrowth, topsoil, rock, animal remains and everything caught in the explosion is bulldozed down the slope, blocking streams and destroying life in the valley below.

By the time of *Goodbye Gauley Mountain*, 500 mountains had been blown up, and an area the size of Delaware cleared. Seen from the air, the scale of devastation appears unreal. Three thousand kilometres of streams have gone, the water system is polluted, neighbouring communities have experienced a 50 per cent rise in cancer rates, black lung disease has doubled and birth defects are now 42 per cent more likely.[x]

The Appalachians are among the oldest mountains on Earth. The range continues the other side of the Atlantic Ocean into the Caledonides, running from Ireland, through Scandinavia, to Hammerfest in arctic Norway. It was born about 300 million years ago, the result of the violent collision of two vast landmasses – Gondwana and Laurentia – which combined to form the most recent of the supercontinents, Pangaea.[xi] The nascent mountains would have been jagged, vicious-looking things – like rows of stony shark teeth eleven kilometres high, stretching across the heart of this mega landmass in the southern hemisphere. It has taken those 300 million years of movement in the Earth's crust, and the erosion and accretion of layers of rock over the course of ice ages and changing sea levels, for the Appalachians to achieve gentle, rolling lushness.[xii]

Coal is readily combustible rock: any stony stuff more than half carbon by weight.[xiii] Anthracite is the aristocracy of the coal world: 95 per cent carbon, glamorous and iridescent. In our store of primary-school facts we remember that coal is plant matter reduced to its carbon by pressure, heat and time. This process is not perpetual: no more coal is being formed. It was a one-off episode, the result of very particular conditions between 360 and 300 million years ago, in warm swampland that sustained fern-like plants, proto trees, weighty amphibians and regal dragonflies. The formation of coal was closely tied to the assembly of Pangaea.

To achieve the right conditions for coal 'you need both a wet tropics and a hole to fill,' explains geobiologist Kevin Boyce. 'We have an ever-wet tropics now, but we don't have a hole to fill. There's only a narrow band in time in Earth's history where you had both . . . and that's the Carboniferous.'[xiv] During the stately collision of landmasses that made up Pangaea, parts of the Earth's crust were rammed upwards (including the Appalachians), but around them troughs and gaps formed. Over 60 million years these holes filled with plant matter, protected by swampy conditions from the bacteria and fungi that would have decomposed it on land. Water levels rose and fell, at times burying the deposits under sedimentary rock and creating thin interlayered seams of what would become coal.

Plants are largely formed of carbon yanked out of atmospheric carbon dioxide. All that vegetable matter buried between layers of rock kept enormous quantities of carbon – and sulphur – out of the atmosphere and cooled the Earth to the point of global glaciation. The coal-fuelled, fire-belching combustion monsters birthed during the Industrial Revolution have exhaled that carbon and sulphur right back out, causing acid rain and global warming. This is not new news. As that great palaeontologist Richard Fortey wrote in 1993: 'Coal burning could be described as the most disastrous exploitation of a geological resource: when time has buried carbon in deep seams, was it wise so to exhume it and consign it to the flames?'[xv]

In 2010 Stephens and Sprinkle married the Appalachian Mountains, pledging to 'speak out, act up, and raise hell about mountaintop removal.' The following year in Asturias – Spain's great mining region – they married coal. Their *Boda Negra Con El Carbón* was messy, sexy, bloody and saturated in coal dust. The audience was 'mainly retired, working class locals', including elderly coal widows. As Stephens and Sprinkle lay naked on stage having lumps of coal dunked in pig blood placed on their bodies, they feared they might get pelted with rotten tomatoes, or worse. Instead the whole community spontaneously formed a circle and sang 'Santa Barbára Bendita' – the plaintive song of Asturian miners.

It's unpleasant being bombarded with environmental statistics and 'home truths'. Ecophilosopher Timothy Morton suggests apocalyptic environmental doom-mongering is counterproductive: people feel disengaged and hopeless, not motivated to push for change. Sprinkle and Stephens's ecosexual performances – funny, shocking, at once knowingly daft and deathly serious – are an engaging alternative methodology, bringing campy humour to climate protest and environmental awareness to Pride marches. 'Why should queers care about climate change?' Sprinkle asks Stephens. 'You can survive without being married, but if you don't have drinking water, you're dead.'

#03

COLTAN

THERE WAS FEVERISH EXCITEMENT AHEAD OF THE LAUNCH OF Sony's PlayStation 2. In Tokyo, queues assembled four days ahead of the console launch on 4 March 2000. Weeks earlier, the PlayStation website had crashed under the rush for pre-orders. Sony sold one million units that first weekend, and a further two million in the five months that followed.[i]

By autumn, when the PS2 was rolled out in North America, Europe and Australia, demand hugely overwhelmed supply. Consumers freaked out as they realised they might not score one of the $299 consoles in time for the holiday season. This side of the story has a happy ending. Gamers came to love the PS2. It remained in production until 2013, shifting 158 million units across the world, becoming the best-selling console of all time.[ii]

There's another side to this story, which is less Grand Theft Auto, more simply grand theft. The year 2000 was a tipping point for the electronics industry. During the 1990s, personal computers and mobile phones had become ubiquitous and their evolution was rapid: products became ever more sophisticated and more elegant. To create smarter, faster, smaller machines, electronics manufacturers rely on rare materials with particular properties well suited to their needs. One is the metal tantalum, used to manufacture small capacitors for hand-held electronic devices.

Tantalum is mined as the streaky black ore columbite-tantalite known as coltan, a mineral that had been rising gently in value until Sony started manufacturing the PS2. Between early 1999 and the start of 2001, the world price of tantalum leapt from US$30 to US$380 per pound (US$66 to US$838 per kilogram).[iii] Coltan is rare stuff, and while it is mined at various sites including Brazil and Australia, in 2000, one major source was the Democratic Republic of Congo.

The DRC was thought to be home to 60 per cent of the world's tantalum resources (this figure has since been revised down[iv]) and the bounce in the value of tantalum caused gold rush-style frenzy, nicknamed coltan fever. In the east of the country, whole communities pivoted to coltan: 'students dropped out of schools; farmers and shepherds left their lands and livestock in favour of artisanal mining activities.'[v] It was a get-rich-quick moment, and many did, though not always those labouring at the rock face.

The DRC's main coltan deposits are in the eastern provinces of North and South Kivu, which border Uganda, Rwanda and Burundi. Kivu's mineral wealth has earned it centuries of bloody unrest, historically fomented by European interests. When coltan fever hit, the region was embroiled in violent conflict, with the government – assisted by Angola, Chad, Kenya, Namibia and Zimbabwe – battling rebels in Kivu backed by Rwanda and Uganda.

In May 2001, the United Nations Security Council produced a report on the exploitation of natural resources and wealth in the DRC. They claimed that this exploitation, by Burundi, Rwanda and Uganda took different, illegal, forms, including confiscation, extraction, forced monopoly and price-fixing. The first two of these had reached proportions that made the war in the DRC a very lucrative business. In other words, these activities were fuelling the conflict.[vi]

Coltan is not Kivu's only treasure. The region is a source of timber as well as tin, cobalt and other valuable minerals, but the coincidental leap in international demand for tantalum was not negligible. 'During the coltan boom, Rwanda even moved prisoners to the Congo, and used them for mining coltan in exchange for reduced sentences and small cash allowances.'[vii] The Second Congolese War earned a memorable nickname: the PlayStation War. Over a decade of conflict, the toll from the fighting and the associated humanitarian crisis was catastrophic, claiming 5.4 million victims: more deaths than any other conflict since the Second World War.[viii]

British artist and director Steve McQueen's short film *Gravesend* (2007) opens with a crucible of seething, molten metal, which cools into a small black disc that reappears in a robotised laboratory. All is clean, cool, depopulated, the only sounds the mechanical whirs and clicks of precision machining. McQueen's camera then takes us into the darkness of a muddy pit where silent men smoke cigarettes and dig through mud to reach bedrock. Bodies labour to extract black-flecked lumps, hitting it from the rockface with crowbars, then tapping rubble with mallets to release the ore. In a tranquil forest stream just beyond, the fragments of coltan are washed and heaped on a pile of large, fleshy leaves. For the mining area of Kivu is also home to the Okapi Wildlife Reserve: this is a beautiful landscape.

McQueen's film takes its title from Gravesend, a Thames port from which the narrator Marlow begins his story in Joseph Conrad's *Heart of Darkness* (1899). The port sits at the start of an interminable waterway that leads Marlow from Europe to Africa, and up another river into the dark heart of violent colonial extraction: the Congo Free State, an absolute monarchy under the personal control of King Leopold II of Belgium. *Gravesend* links the scramble for coltan to centuries of violence and exploitation, the colonial states of centuries past replaced by multinational electronics manufacturers.

Trying to uncouple coltan mining from regional conflict, the international community has since imposed boycotts and regulations on electronics manufacturers, miners, exporters and the DRC itself. In 2010, a clause in the Dodd-Frank Act (a reform of financial regulations) forced US firms to audit their supply chains and ensure they were not using minerals from a source that funded Congo's war. Mines closed for six months while the government got a handle on the new restrictions. Even afterwards, manufacturers like Apple and Intel steered clear for fear of bad press.[ix]

In practice, legislation has been near impossible to impose. The DRC has coltan, and electronics manufacturers need it. Vast amounts of the mineral are smuggled into neighbouring countries: corporations instead purchase ore 'mined' in Rwanda. In theory, sacks of coltan are meant to be tagged and traceable: in 2015, the head of North Kivu's mining division informed a reporter that because of the prohibitive cost, not a single mine was tagging its output.[x]

Mines are inspected for child labour and unsafe practices, but the territory is vast and infrastructure poor. In 2021, a report by Oluwole Ojewale from the Institute for Security Studies detailed the still-growing demand for Congolese coltan, much of which was extracted using the 'labour of over 40,000 child and teenage miners'. These children face dangerous mining conditions, criminal activity, harassment, abuse and the hazard of daily exposure to the radioactive substance radon. Manifestly, attempts to control the trade are failing.[xi]

Many reporting on coltan mining have come away from the DRC with a different story from the one they anticipated. The industry brings money into an area with low employment, and boycotts and bans destroy jobs. 'It's relatively easy to source minerals in a warzone and pour them into an international market that demands ever-niftier gadgets,' writes Tom Burgis. 'But regulating the supply chains of our global economy – without inflicting harm on whole communities by choking off livelihoods in far off lands – is an altogether harder task.'[xii]

#04

FLINT

F LINT JACK WAS BRIEFLY NOTORIOUS AS THE 'PRINCE OF Counterfeiters' a man sunk deep in the mire of rascality, and a celebrated rogue.[i] Uncredited until recently, his work is present in the collection of almost every museum of natural history or antiquities in England, Scotland and Ireland.[ii]

Unsurprisingly for a lifelong counterfeiter, he was an unreliable narrator of his own life. In different locales he was known as 'Old Antiquarian', 'Skin & Grief', 'Fossil Willy', 'Snake Billy', 'Bag o'Bones', 'Shirtless' and 'Cockney Bill'.[iii] His real name was understood to be Edward Simpson, though at various different times he went by John Wilson, Jerry Taylor, Edward Smith and Edward Jackson. He was born at Sleights outside Whitby, or in Carlisle, or possibly 'the city of Derry, Ireland, brought from the "dear country" to Scotland by his stepfather at the age of eight years', though many who crossed his path suggested his accent was Cockney, and London his most likely origin. [iv]

In the version of his biography that appeared in Charles Dickens's magazine *All the Year Round*, Flint Jack starts life as an honest fellow, at the age of fourteen, entering the service of Dr George Young in Whitby. Young was a historian, and contributed much to Whitby's reputation as a site of geological and paleontological significance. In the early nineteenth century, Whitby, together with Lyme Regis on the south coast, had become the focus of a mania for fossil hunting and collecting (see also: Blue Lias, p. 262), and during Flint Jack's period of service, discoveries included a partial plesiosaurus skeleton. The five years Flint Jack spent working with Young would have furnished him with an intimate knowledge of the rocks of the Yorkshire coast and the treasures to be found in them.

He then spent six years with Dr Ripley, co-owner of the Whitby Museum, and like Dr Young, a fossil hunter.[v] Upon Ripley's death our hero, now twenty-five, embarked on his roving life. Thanks to the efforts of Dr Young and others, a fine supply of fossil-hunting tourists passed through the area, offering a ready market for specimens that turned up in the shale of the cliffs.

By the mid-nineteenth century, fossil collecting was giving way to a Victorian fascination with anthropology.[vi] Among Danish archaeologists, new ideas were emerging about human prehistory, and the sequence of developmental progressions that had taken place en route to the perfected specimen: the rational European man of the day. In 1849, Jens Jacob Asmussen Worsaae's theory of the three-age system – categorising early human cultures according to their use of stone, bronze and iron tools – became available in English translation. For centuries the discovery of stone axe-heads and flint tools had been explained according to popular folklore: large stone implements were 'thunderstones' driven into the earth in a violent lightning strike; arrow-heads might be 'elf-shot' fired by fairies to injure cattle. Long collected as curiosities, flint tools were now prized evidence of the earliest Britons.

Knapping requires skill: angled flakes of flint are knocked off with a hammer stone or bone. The fine flakes were used as cutting implements, sharp enough to slice animal hides and butcher carcasses, while the main stone was formed into a heavier implement. By the Neolithic period (c. 4000–2300 BCE) and the shift from hunter-gatherer societies into early agriculture, the flint knapper became a specialist role. Flint was mined at sites such as Grime's Graves in Norfolk, and well-knapped tools used in trade.

It is not clear how or when Flint Jack realised it was faster to manufacture flint tools than scour the fields for them. Although he had the advantage of access to a metal hammer and nails, the skill required to produce convincing stone arrow-heads should not be underestimated. At the height of his output, he claimed to have been able to manufacture and sell fifty arrow-heads in a day, walking as much as fifty kilometres and peddling his 'finds' to those he met on the road.

He sometimes diversified – rough-firing local clay to look like ancient burial urns; hammering a metal tray into a Roman-style breastplate; inscribing cod Latin into a stone that he trundled into town in a wheelbarrow; carving fossils into rock fragments; stealing an elephant's tooth, burying it in mud, then trying to pass it off as mammoth ivory – but it was his work with flint that saw the greatest expression of his skills, and of his imagination. No longer content with the repertoire of early tools – arrow-heads, spear tips, blades, scrapers, burins, axes and so forth – he improvised, producing fish-hooks and combs. So distinctive was his work, that local historian Thomas Wright, upon studying collections amply furnished by Flint Jack, concluded that they 'featured artefacts of a distinct culture, unique to North and East Yorkshire, identifiable with the Parisi, a Celtic tribe described by Ptolemy.'[vii] Wright was greatly embarrassed when Edward Hawkins, a curator at the British Museum, suggested his entire thesis was based on the study of forgeries.

Hawkins was not the only expert to rumble Flint Jack: collectors were becoming aware of the parallel trade in counterfeits, and displays were mounted in regional museums to help the public identify faked artefacts. A more significant setback, in 1846, was Flint Jack's turning to drink: 'the worst job yet. Till then, I was always possessed of five pounds. I have since then been in utter poverty, and frequently in great misery and want.'[viii]

His mounting notoriety among collectors may have blocked the path to the sale of counterfeits, but it generated an audience hungry to hear of his exploits. In 1866, the Christmas edition of the *Malton Messenger* – one of two weekly newspapers in circulation in this small North Yorkshire town – opted not to publish a sentimental festive tale and instead dedicated its pages to the life of Flint Jack.[ix] The edition quickly sold out, and the story was picked up by the national press.[x] The forger was photographed by postcard sellers; archaeologists would hang a portrait of him in their offices, framed in his own flints; and he even gave a demonstration of his skills to a meeting of the Geological and Archaeological societies in London. Flint Jack's champions – and he had many, even among those he had duped – proposed that he might make an honest career producing artefacts to order, but he remained enamoured of counterfeit.

In January of 1867 he visited his old customer James Wyatt in Bedford, who reported that he was 'very poorly clad, unshorn and shivering with cold and hunger'.[xi] Wyatt fed him, gave him clothes and money for lodgings and commissioned a set of flint tools for a display, but Flint Jack was in too disorderly a state to fulfil it. He drank his way through the available funds, then returned to Wyatt's village and made two clumsy attempts at burglary. After peddling a clock from a Methodist chapel to a publican, he was picked up by the police. It was only with the intercession of Wyatt, who testified that his character was not beyond redemption, that Flint Jack was sentenced to a year's incarceration rather than transportation to a penal colony.

Little is known of Flint Jack after prison. The last record of his appearance is at a magistrate's court in Malton in 1874. There are rumours that he died in a workhouse. *All the Year Round* ends his story on a moralising note: 'But what a waste of ability! What might not this man had done for science had he only taken the same pains in assisting as he did in leading it astray!'[xii]

Perhaps Flint Jack might instead be celebrated as an anarchic force. In a period when men of science came to believe humankind was growing ever more glorious 'as if the Creator's skill had improved by practice',[xiii] how satisfying it must have been to hoodwink them with Stone Age skills.

HAÜYNE

THE PHYSICIST AND BOTANIST ABBÉ RENÉ-JUST HAÜY WAS a lover of order, condemned to live in very disordered times. A humble, otherworldly man of intense focus, it seems he paid scant attention to the rapid political changes unfolding in Paris during the summer of 1792. On 10 August, armed revolutionaries stormed the Tuileries Palace and King Louis XVI and his family were imprisoned in the Temple. A few days later, Haüy was apparently quite startled to have the bookish tranquillity of his modest rooms at Cardinal Lemoine College upset by hostile visitors. On being asked whether he had firearms on the premises, Haüy responded playfully, sparking off a piece of electrical equipment with which he had been experimenting. This theatrical gesture may have lightened the mood, but he was arrested nonetheless, for being both a cleric and a member of the Academy, and for failing to take an oath of fidelity to France that recognised its supremacy over the Church. The algebra covering his papers was suspected of being a cypher concealing subversive messages, and all were seized. In a still greater blow – from Haüy's perspective – his interrogators messed up his treasured collection of mineralogical samples.[i]

Haüy was imprisoned at the nearby seminary of Saint Formin along with other priests who had rejected the oath of loyalty of state over Church. Despite the very real danger of his position, he seems not to have minded the change in location, from one modest cell to another, particularly after his jailers reunited him with his drawers of mineral specimens. He busied himself re-organising them.

These were volatile months. While Haüy may have been oblivious of the danger, his former colleagues from the Academy and the Jardin du Roi[ii] were aware of the violent hostility mounting against recalcitrant clergy. It is a testament to their affection and respect for Haüy that they were prepared to endanger themselves by pressing for his release. They were successful, and the naturalist Geoffroy Saint-Hilaire rushed to the prison only to find Haüy engrossed in his work and refusing to move at such a late hour. He was again reluctant the following morning, and Saint-Hilaire could only induce him to leave the prison after considerable persuasion.[iii] It was just in time. Only a few days later, on 2 September 1792, eighty of the priests who had been incarcerated with Haüy at Saint Formin were killed. In the bloody Massacres du Septembre that followed, 1,200 Catholic prisoners were killed.

During the Reign of Terror, the Royal Academy of Sciences – and indeed all academies – had been dissolved. In May of that year, the eminent chemist Antoine-Laurent de Lavoisier, who had recognised and named oxygen and hydrogen, was executed by guillotine. Yet despite the initial mistrust of the Republic, in the conflict that followed scientists became an important asset: they were needed to produce hydrogen for balloons, turn church bells into canons, and manufacture saltpetre for explosives.[iv]

Due to his evident unsuitability for martial matters, Haüy was left largely to himself until his services were called on to fill the post of professor of physics at the École Normale. In 1794, the greatest scientists and academics in France were brought together and instructed to train a new generation of secondary-school teachers in the most advanced ideas of their time. Naming a training academy run by a professorial elite the École Normale (Normal School) was a political proclamation in itself. At the behest of the National Convention, education was to be standardised, just as systems of measurements were. The knowledge of the elite was to become the knowledge of the masses.[v]

A combination of theatrical flair and down-to-earth integrity made Haüy a popular teacher at the École Normale. He performed spectacular experiments with electricity and magnetism, but then picked them apart for his students. This, he explained, was the difference between a charlatan and a physicist: the first hides the mechanism and professes himself owner of a magical art, while the second reveals and explains.[vi]

While he was a scholar of physics and botany, Abbé Haüy is most remembered for discovering the law of crystallisation.[vii] It is a field that appealed to his love of order twice over: by studying the regular internal structure of crystals he was able to establish a system of classification for minerals. Assisting a colleague at a lecture in the Jardin du Roi he observed a dropped specimen of calcareous spar breaking along clean lines and leaving regular fragments. (Legend has it that it was Haüy himself who dropped the specimen. A rival apparently took great pleasure in referring to him as a 'cristalloclaste' – a breaker of crystals.[viii]) 'The prism had a single fracture along one of the edges of the base, by which it had been attached to the rest of the group,' Haüy wrote. 'I tried to divide it in other directions, and I succeeded, after several attempts, in extracting its rhomboid nucleus. This at once surprised me, and gave me the hope that I could advance beyond this first step.'[ix]

Haüy was the first to develop the important fundamental notion that every crystal is built up by an orderly stacking of tiny sub-units, and that the final shape of the crystal depends both on the shape of these small units and the way they are packed together.[x] His *Traité de minéralogie* (Treaty on Mineralogy, 1801) was known and taught around Europe. Haüy also introduced beautifully carved pearwood models as a three-dimensional teaching aid, collections of which still exist. A set would have run to about 600 pieces, each in the crystal form of a different mineral analysed and identified by Haüy.

Perhaps it was its crystalline structure that inspired the Danish scholar Tønnes Christian Bruun-Neergaard to dedicate a newly discovered silicate mineral to Abbé Haüy in 1807. It was unusual, at the time, for minerals to be named after people, but Haüy had a transformative impact on the field. 'The science of crystallography was entirely created by Haüy's genius, and his successors have had little to do except to perfect the details of his work,' opined one of those successors, the nineteenth-century mineralogist François Mallard. 'No other branch of human knowledge is in the same degree the work of one man alone.'

Haüyne, the rare mineral named in his honour, occurs in a zippy blue that declares modernity. It's a colour you might select to communicate cleanliness and efficiency: the blue of a powerful detergent, sports drink or electricity. It's a flashy stone that sits oddly with Haüy's monkish asceticism and preference for drab clerical attire (he notoriously objected to wearing ceremonial robes when elected to the Royal Academy of Sciences in 1783).

Haüyne is a rather brittle mineral of the sodalite group, and has what mineralogists call 'perfect cleavage' – meaning not that said crystal has a bright future as a swimwear model, but that it cleaves with a smooth surface, the characteristic which first caught Haüy's eye, and led him to investigate mineral structures.

After Napoleon came to power in 1799, Haüy was appointed professor of mineralogy in the Muséum national d'Histoire naturelle. Napoleon, in his new role as First Consul, commissioned him to write a treatise on physics to be used in all French schools. It took the enforced leisure of exile on the island of Elba in 1814 for Napoleon to find time to read the book he had commissioned thirteen years previously. It made an impression: on his return to Paris in 1815, he appointed Haüy chevalier de la Légion d'honneur.[xi] The order that Haüy found in the crystal world being quite absent from the French political scene at that time, it was a rank that he only enjoyed for a few days, until Napoleon's defeat at Waterloo and the restoration of the monarchy that summer.

#06

LODESTONE

L ODESTONE IS THE LOVER OF THE MINERAL WORLD. IT LONGS TO couple, either with its kin or its cousin iron. Lodestone is a natural magnet, occurring in an oxide of iron called magnetite, named by the Greeks after Magnesia in Asia Minor. Not all magnetite is inherently magnetic, ergo, not all is lodestone. As with the static electricity produced by amber, early writers found in lodestone's uncanny powers of attraction proof of matter's vitality. Having described lodestone, Theophrastus speculates that certain stones might 'give birth to young'.[i]

Clues to lodestone's amorous nature endure in the French for magnet – *aimant* – loving, affectionate. The fourth-century poet Claudian offers a frankly raunchy description of the mating of two devotional statues in a temple, one a polished iron figure of Mars, the other a lodestone Venus, who draws her lover to her with powerful ardour: 'The stone, afire, sighing, smitten, perceives congenial material/ and the iron finds quiet attachment.'[ii]

In the eleventh century, Marbod, the bishop of Rennes, suggested that any man wishing to test their spouse's virtue should place a lodestone beneath her head while she sleeps: if chaste she will roll over and embrace her husband, but if false she will tumble out of bed as though repelled.[iii] Modern magic advises feeding a lodestone with iron filings to keep it potent, working with a single stone to attract power, favours, money and gifts, and paired lodestones for love.[iv]

Lodestone compasses are attracted not to love, but to the Earth's magnetic poles, although their origins are spiritual rather than navigational. Developed during the Han dynasty (c. 200 BCE–220 CE), the earliest compass was a hemispheric bowl of lodestone with a little tapered handle, like a stubby ladle, which pivoted south. The 'south pointing spoon' was placed bowl-down on a polished bronze circle at the centre of a chart inscribed with concentric rings of symbols and systems. All related to the cosmos and the natural world: the constellations; the eight trigrams of the *I Ching* (Heaven, Lake, Fire, Thunder, Wind, Water, Mountain, Earth); and twenty-four cardinal directions. It was a divinatory tool, used in Feng Shui to determine the most auspicious orientation for a building, position for a tomb, or time and place for an important event.[v]

Only in the twelfth century did Chinese sailors use the compass for navigation, an innovation that facilitated maritime trade between China and the Islamic world.[vi] Compasses had become lighter and more portable, using iron needles magnetised with lodestone inserted into wooden fish set to float in a bowl of water (a 'south-pointing fish'), or suspended from silk thread. The compass ranks alongside paper, printing and gunpowder as one of the 'Four Great Inventions' of ancient China, but whether it was an invention gifted to the West, or whether European scientists developed a navigational compass around the same time is open to debate.

The Chinese compass very probably arrived in the Mediterranean from the Islamic world, but that did not prevent stories of its having originated in Europe from flourishing. On the Amalfi Coast is a statue of mariner 'Flavio' Gioia, honouring him as the inventor of the compass in 1300. Whether or not Gioia actually existed (the compass needle of research currently hovers at 'no'), his supposed invention had by then been in use in Europe for decades.[vii] Over a century earlier, the English scientist Alexander Neckam in his *De Naturis Rerum* (1187) had described sailors navigating with the aid of a magnetised needle.[viii]

The polyglot Mediterranean's compass needle pointed north, not south, and its rose was marked with the directions of that sea's eight winds – Tramontana to the north, then Greco, Levante, Scirocco, Ostro, Libeccio, Ponente and Maestro. The fleur-de-lys drawn at the top of the compass rose in the fifteenth century by Portuguese cartographer Pedro Reinel is the stylised 'T' of Tramontana. By then, navigators had realised that there was a discrepancy between magnetic and 'true' north that needed to be factored in as they travelled further from the equator.

While great ocean-going peoples – the Vikings in the Atlantic, and the Polynesians in the Pacific – had navigated long voyages by the waves and the heavens, the arrival of the compass allowed other seafaring nations to travel beyond sight of land, and in conditions where the sun or stars were obscured. In Europe, the arrival and adaptation of the maritime compass opened the seas for the age of sail, and of global exploration (see also: Alunite, p. 16; Ruby, p. 49). Trade and warfare could be conducted almost year-round, and strange waters charted: movement across the Mediterranean increased, as did sea traffic beyond the straits of Gibraltar. 'Lodestone' dates from the sixteenth century and means 'way' or 'journey' stone. The vitality observed by Theophrastus had become a dynamic force.

#07

MICA

For the artist Ilana Halperin growing up in New York City, mica brought something like magic to the surface of her surroundings. 'I knew mica from streets that glinted in the sun, playgrounds peopled by boulders that seemed made of silver and gold, rocks on the beach with layers you could peel open like pages in a book,' Halperin wrote on her geological memory-map of the city, *Minerals of New York*. Origins were important: Halperin's grandmother had arrived in New York after fleeing pogroms in Lithuania. History is bound up in bedrock: 'A mineralogy curator named Peter told me mineral samples of mica are sometimes termed "books". My mother remembers finding books of mica in the alley next to the building where she grew up in Brooklyn.'[i]

Shimmering silicate flakes of mica give granite its disconcerting glamour. It sparkles in Manhattan schist, the island's ancient metamorphic bedrock. A common variety known as muscovite – also 'cat silver' – has the translucent lustre of eroded seashells reduced to their mother of pearl. Prised apart into thin sheets, in southern India it was painted to provide decorative covers for temple lanterns, and in Russia and the US, was used in window panels as 'Muscovy Glass' or isinglass. Curly McLain's carriage in *Oklahoma!*, fitted 'With isinglass curtains y' can roll right down/ In case there's a change in the weather'[ii] would have had a little peephole fitted with mica: suitable for a vehicle, because much lighter than glass and less likely to shatter.

Resistant to extreme temperatures and non-conductive, Muscovite mica was used in thin sheets as stove windows and as fine insulating layers in early electrical devices. During the Second World War, its use in radios and communication equipment made it 'most essential to the successful prosecution of this war of machines containing electrical units'. The conflict cut European countries off from processed mica imports from India, putting the mineral 'at the top of the list of most critically needed materials'.[iii] The Geological Survey of Great Britain identified deposits in the pegmatite at Knoydart, a remote district on the west coast of Scotland, where mica was mined and rough-dressed before being driven inland to Pitlochry for the delicate job of separating the books into fine sheets.[iv]

During the same period, in what is today the Czech Republic, mica was worked at the Theresienstadt ghetto. Supposedly an example of the new, self-contained communities being created for Jews deported from around Europe, for a brief period in 1944 the ghetto was beautified, used to hoodwink a visiting delegation from the Red Cross and to create a propaganda film showing the luxurious life of 'parasite' Jews. As German newsreel commentary put it: 'While Jews in Theresienstadt sit and dance over coffee and cake, our soldiers carry all the burdens of a terrible war, hardship and privations, in order to defend their homeland.'[v] In reality it was a forced labour camp, a transit station for Auschwitz,[vi] and place of slow death.[vii] From June 1942 thousands of Jewish women and girls spent ten or twelve hours a day at the *Glimmerwerke* (mica works) splitting sheets apart with razorblades for the Reich Office for Electrotechnical Products, the ghetto's largest military contract.[viii]

The women of the *Glimmerwerke* had daily quotas by weight, and were forced to work until those were met, leaving them at risk of missing food rations if they were too slow, or slicing their fingers if they were hasty. 'Our life hangs on the tongue of the scale,' wrote Emma Jonas, who carried fragments of mica from the *Glimmerwerke* with her throughout sixteen years as a displaced person after Theresienstadt was liberated by Red Army troops on 8 May 1945.[ix]

Emma Jonas followed earlier waves of Jewish migrants, finally arriving in the United States in 1961. During the many years of her journey, with the fragments of mica tucked in her purse, the use of a miraculous new plasticised material developed to take the place of mica in electrical products for the military had become mainstream. This was Formica, by 1961 a fixture in US homes as an affordable and hygienic kitchen surface.

#08

MILLSTONE
GRIT

T HE NAME EVOKES SUCH REMORSELESS, ABRASIVE HARDNESS AS TO sound almost like parody: millstone grit. A moniker fit for the wrestling-ring. And it does endure: on Ilkley Moor, great staunch purplish cliffs of the stuff stick out above seas of bracken and heather. The wind up there is cruel, slicing through layered wool and whipping words away unheard. Over centuries, wind and rain have sculpted the stones, but they are still rough as emery.

Formed from silt deposited in a great delta during the Carboniferous period (see: Coal, p. 223), millstone grit is tough sandstone cemented with quartz.[i] Particles of ancient granite, eroded then built up as waters of a tropical ocean retreated from what is now the east side of the (distinctly untropical) Pennines in the north of England. It has been quarried for grand buildings – including the magnificent Chatsworth House – but for thousands of years its abrasive weight has also been used to process grain.

'Since no important civilization has been developed without grain, the mill has ever been the handmaiden of material culture,' wrote agricultural historian Russell H. Anderson in 1938.[ii] The earliest grain stores – 11,300 years old – have been found at the Neolithic site of Dhra' near the Dead Sea in Jordan.[iii] Constructed before the earliest domestication of plants (in other words, before the shift from hunter-gatherer to agricultural societies) the granaries were equipped for the storage and processing of barley, with grinding stones (querns) built into the structure. Querns – also known as saddle stones – have a shallow dip to contain grain that is ground with a hand-held stone. Portable querns were used by early cultures from Central America to Central Asia. They are the millstones of the Old Testament world.

The Romans developed the quern into a machine that could be operated by two women, with a domed stone beneath and moving millstone above. Grist is poured in through a hole at the top, and meal appears at the edges. The size of a mill was limited by the power available to drive it: centralised bakeries for large communities needed a more substantial energy source.

The thirty-one mills found preserved at Pompeii were large and heavy – pumice querns domed like beehives, with bell-shaped mill-stones above – and would have been turned by animals and enslaved people.

Hydraulic technology came to the aid of ground-down human and animal labour. In *De architectura* (c. 30–15 BCE) the Roman military engineer Vitruvius describes a machine that transmits the power of a vertical waterwheel to drive horizontal grinding stones: a design that endured without substantial improvement for about eighteen centuries.[iv] Archaeological remains of early water mills are rare – they were located on wet and unstable territory, and stones from old buildings tend to be reused in new ones – but the handful found in Greece date back to as early as the fifth century.[v]

The Romans introduced watermill technology to Britain – though millstone grit also furnished querns for home milling for centuries before and after. Today there are so many thousands of millstones – used and unfinished – strewn about the East Pennines that the Peak District National Park adopted one as its emblem. They are cumbersome, unshiftable things, yet must be refined with grooves of proper depth, and skilfully balanced to sit evenly one upon the other. When white bread came into fashion, gritstone milling went out: the flour was too coarse, grey and sandy. Finer millstones took their place, and demand for the tough and gritty stuff abruptly ceased. (The grindstone to which you put your nose is not a mill but a turning wheel for metal work: finer sections of gritstone destined for the steelworkers of Sheffield.)

Dark, brooding millstone grit lives on in the emotional mythology of the moors and peaks. This is a disorienting landscape of low scrubby heather, bilberries and bog, populated by speckle-faced sheep, chuckling grouse and curlew. It's easy to lose yourself here, body and mind. In her foreword to *Wuthering Heights* (1847), written a year after her sister Emily's death, Charlotte Brontë describes the novel 'hewn in a wild workshop, with simple tools, out of homely materials'. Emily is imagined as a sculptor chiselling a block of stone discovered standing on the moor. 'With time and labour, the crag took human shape; and there it stands colossal, dark, and frowning, half statue, half rock: in the former sense, terrible and goblin-like; in the latter, almost beautiful, for its colouring is of mellow grey, and moorland moss clothes it;

and heath, with its blooming bells and balmy fragrance, grows faithfully close to the giant's foot.'[vi]

In describing her sister's labours, Charlotte cleaves to the geological lexicon of the book itself. Heathcliff's very name is built from the scrub and rock of the moors, while Catherine's love for him 'resembles the eternal rocks beneath – a source of little visible delight, but necessary'.

One hundred and fifty years later, that 'necessary' binds Anne Carson's poem 'The Glass Essay' like quartz in gritstone. Her narrator, grieving for lost love, reads Emily Brontë as she travels to stay with her mother on the moors of an unspecified 'north'. She imagines that the relentless grinding of wind on the millstone grit of the moors 'taught Emily all she knew about love and its necessities—/ an angry education that shapes the way her characters/ use one another.'[vii]

The obstinate, abrasive stone becomes the texture of discomfort between people: her mother nurturing arguments; her father's memory fragmented by Alzheimer's; the furious, heartsick, clawing sex of a relationship already passed.

> *Soul is the place,*
> *stretched like a surface of millstone grit between body and mind,*
> *where such necessity grinds itself out.*

09

OBSIDIAN

I N THE WORLD OF MINECRAFT, OBSIDIAN IS ONE OF THE MOST PRIZED practical resources. The video game presents obsidian as a material of exceptional strength, able to withstand explosive blasts and invaluable in structural defences. Formed by pouring water onto lava, Minecraft's obsidian must be mined with a diamond pick-axe, and is so resilient that even the mighty Ender Dragon cannot destroy it. It is most prized for qualities that transcend its physical strength. Obsidian has supernatural powers, with which players can construct a portal to the Nether, a sulphurous underworld laced with flowing lava governed by parallel rules of time and space.

In 2019, Amy L. Covell-Murthy from the Carnegie Museum of Natural History welcomed a local Scout group training for their merit badges in archaeology. The Scouts undertook excavations in dig boxes, and arranged and labelled stone tools – arrow-heads and spear tips made of flint and obsidian – into small museum exhibits. After a spot of curatorial myth busting, one Scout's caption, appended to a grizzly, notched obsidian spearhead, ran: 'Minecraft lied to me! Obsidian is not unbreakable! Obsidian cannot be used to make portals! It's volcanic glass!'[i]

Outside Minecraft, you won't create obsidian by pouring water onto lava, but it does form after a rapid drop in temperature. High-silica lava solidifies into glass when it cools too fast for visible crystals to form. Typically black, obsidian has been found in many colours (though not the purple bestowed on it by Minecraft). It doesn't have the structural strength of rocks such as granite, but flakes like chert or flint if hit at the correct angle (see: Flint, p. 233). It is indeed a prized resource, or has been, historically. Wherever in the world obsidian occurs, people have used it for cutting tools: blades, scrapers, spears and arrow-heads.

In Jōmon era Japan (c. 10,000–300 BCE) obsidian was mined at sites across the archipelago. The finest came from the mountainous Nagano Prefecture of Honshū. Here, amid the trees of the Hoshigatō archaeological site, 193 obsidian mine pits have been excavated. Remains of ceramic objects and tools suggest the pits were worked for millennia. Mining methods changed between the Early and Final Jōmon: a shift from excavation with deer antler picks, to the use of hammer stones to chip obsidian from its rocky source.[ii] Even in the Early Jōmon period, obsidian from the site was traded across a broad territory, from the Tōhoku to the Tōkai and Hokuriku regions. Hoshigatō carries the glossy stone in its name: the character 'Hoshi' means 'star', suggesting the sparkling glass had fallen from the sky.

A fine flake of obsidian can be extraordinarily sharp: in the late 1990s, a surgeon at the University of Michigan attracted national coverage after revealing to a campus magazine that he used medical-grade obsidian blades for cosmetically sensitive procedures such as mole removal or earlobe repairs. 'I like the obsidian knife because it traumatizes the tissue less,' Dr Lee A. Green told the *University Record*. 'It is very sharp and very smooth at the microscopic level.'[iii]

Covell-Murthy's story about the Minecraft-playing Scout is part of a long-running debate about the educational benefits of the mining and construction-oriented game. In the years since Markus 'Notch' Persson launched it in 2011, academics have written about Minecraft's importance as players' first meaningful exposure to powerful ideas in STEM (Science, Technology, Engineering and Mathematics) subjects.[iv] Pedagogues have praised its potential for engaging children alienated by classroom learning environments. Tech journalists have noted how gameplay encourages 'computational thinking'.[v]

The Hudson Institute of Mineralogy's Minedat.org blog lists mineralogical counterparts of the game's fictional substances such as prismarine, glowstone and netherrack, but notes: 'As far as we know there are no natural forms of obsidian that cry.'

'When a parent says, "I hear it's educational," I imagine they are actually thinking, "I hear it's a video game that I can let my kid play and not feel guilty,"' quips author and technologist Hana Schank.[vi] It's understandable: in the face of Minecraft's overwhelming popularity (141 million active players worldwide and counting) it helps to feel it's a force for good.[vii] Minecraft play isn't all STEM subjects, though: it's also about magic, monsters and worldbuilding.

Throughout history, the otherworldly, terrifyingly sharp material obsidian has been endowed with supernatural qualities.

Native Americans of the volcanic west coast – among them people of the Paiute, Shasta, Wintu, Yurok, Yuki, Achomawi and Pomo nations – have been custodians of a deep knowledge of obsidian. They understood how to work it, the suitability of different sources for various types of tool, and the least damaging ways to obtain it. Obsidian plays a role in their earliest stories, their histories and cautionary tales – manifesting in rivers, falling from the sky and bestowing special powers. Buried in these myths and tales are actual eyewitness accounts of the fiery formation of obsidian at sites such as Glass Mountain, oral records of volcanic activity in the deep past.

At the turn of the twentieth century, one of the Yuki people described the great spirit Milili who 'owns an enormous block of obsidian, of which all obsidians in the world are fragments that he has thrown down. He has the shape of an enormous eagle or condor and controls deer, *mil*, to which his name refers. *Kichil-lamshimi* – obsidian doctors – *mil-lamshimi* – deer doctors – and *mit-lamshimi* – sky doctors . . . all derive their power from Milili.'[viii]

Spirits can shoot invisible obsidian points into the body: in extracting them, obsidian shaman cured otherwise unexplained ailments. They kept great obsidian blades, some of the longest in the world, stretching to almost a metre, which were used only for ceremonial purposes, and passed down through generations.

A Yuki shaman describing a ritual experienced in his youth, as his powers started to emerge, recalled dreaming of the creator On-uha"k-namlikiat and finding himself in the sky where he 'saw many colours, like a mass of flowers. In the morning I was bleeding from the mouth and nose and badly frightened.'[ix] His powers grow over the course of many dreams, until he at last dreams of the obsidian spirits 'but did not reply to the spirits who were addressing me, thinking the dream would come to me again, but clearer. Later I was told by the old people that I had made a mistake – that the obsidian spirits never spoke to anyone more than once.'[x]

As a result, he could cure all diseases save for those caused by spirit obsidian.

Considering the importance of obsidian to myth and ritual, the Minecraft-playing Scout's caption bears revisiting. Can obsidian open portals to other realms? That depends which world you're in, and who you ask.

LIVING
STONES

BLUE LIAS

CALCULI

COPROLITE

CORAL

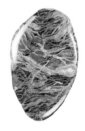

LEWISIAN GNEISS

PEARL

SLATE

SULPHUR

THE UJARAALUK UNIT

LIVING
STONES

STONE HAS SURETY – IT'S THERE IN OUR IDIOMS, SUPPORTING language like bedrock. Something unchangeable is set in stone; someone beyond reviving is stone dead; you can be rock solid, and rock steady, but if you lack human empathy you are stone hearted and flinty. Stone is imagined as eternal, like the torment of Sisyphus, pushing his boulder. We are taught that life is short, ergo this durable stuff has come to represent its antithesis.

This is a matter of perspective, or, more accurately, time. Stone moves, grows and dies. Continents have traversed the globe, and their journey continues: the tectonic plates of North America and Eurasia are moving away from each other. Wet ocean crust is being subducted beneath continental plates. Mountains are eroded by high winds and rain. Volcanic activity is building islands. New layers of sediment accrue in the oceans. Observed in mountain time – rather than human time – there's a lot going on.

Living entities of all kinds are bound up in tight symbiosis with their stone environment: witness the huge loss in biodiversity that occurs when coastal reefs are replaced by concrete barriers.[i] Stone is not always slow, neither is it necessarily a thing apart from the living body (think how fast a human gallbladder can manufacture concretions). We have the hard matter generated by ancient life forms to thank for marble, chalk and limestone. In the face of a complex entity like coral, the categories posed by the guessing game 'Animal, Vegetable, Mineral?' fall short.

Stone has played a huge role in nurturing life, furnishing habitats, nutrients, natural barriers and tools. In 2008, a team led by

Robert Hazen presented a radical new proposal: that life, conversely, has played an extraordinary role in the evolution of stone.[ii] The theory of Mineral Evolution charts growing diversity and complexity over more than 4.5 billion years. A dozen 'ur-minerals' made up the matter that formed Earth: today 5,100 mineral species are known.

The flourishing of life on our planet assisted that diversity, most dramatically during the Great Oxygenation Event caused by cyanobacteria developing photosynthesis over 2.5 billion years ago. Atmospheric oxygen rose, and with it came an extraordinary proliferation of new mineral species – from about 1,500 to over 4,000 varieties.[iii]

Life – specifically human life – continues to have an enormous impact on the mineral world. Our plastic debris and other durable ephemera will form the fossils of future Earth. Hazen's team has identified 208 new minerals that have developed naturally as the result of human activity (in mines for example), as well many hundreds more mineral-like manufactured compounds including cement, ceramics, batteries and cellphone components. The proliferation of new mineral and mineral-like species is one of the most striking transformations taking place in the anthropocene, and it is occurring in human, rather than mountain, time.

#01

BLUE
LIAS

THE LIMESTONE AND SHALE FORMATION KNOWN AS BLUE LIAS IS not so much blue as the ominous greasy colour of soft pencil leads. Formed from silt deposited in the mucky seas of the Jurassic, between 195 and 210 million years ago, it's an oceanic layer cake of dead matter that looks decidedly inedible. Actually it's deadly stuff: huge slumps of freshly descended clifftop sit on the beach, spewing out limbs and roots of spindly trees above a tumble of oily rock that splits along tight sheets. The Blue Lias is perilously unstable, casting itself seaward at a rate of metres per year.

There's nothing mystic about the name, it's a locational quirk – 'lias' a phonetic rendering of 'layers' as described by Devon quarrymen in the early years of British geology.[i] With Blue Lias too soft and easily weathered to be a good building stone, that quarrying was largely for limestone to make lime and cement.[ii] Yet the Blue Lias formation, found near the small town of Lyme Regis in the South West of England, has a romantic hold on the geological – and paleontological – imagination.[iii] It's the rock of seekers and dreamers, carrying the promise of fossilised fish, spear-tip belemnites and the horny bivalve *Gryphaea arcuata*, known evocatively as devil's toenail. Embedded in the larger boulders of the beach are ghostly traces of ammonites as big as cartwheels, and smaller specimens in jagged profusion, their traces jumbled into abstraction like a lithic Kandinsky.

In the early nineteenth century the crumbling Blue Lias yielded a treasury of ancient reptiles – ichthyosaurus, plesiosaurus, pterodactyl – to a European public with an insatiable greed for collecting, naming, painting and cataloguing all aspects of their world. The finest, and often the first, of these specimens were coaxed from their Lias deathbeds by Mary Anning, a celebrated figure, then as now.

Humans have found fossils for as long as the Earth has eroded, crumbled and yielded past treasures, but they were explained in other terms: thunderbolts, snakestones, toads in the rock. In the fifth century St Augustine, finding an enormous tooth on the shore at Utica, saw it as evidence of ancient giants: 'the bones which are from time to time discovered prove the size of the bodies of the ancients, and will do so to future ages, for they are slow to decay.'[iv] Anning's finds were presented to a very different public, one that was starting to understand fossil objects as former occupants of Earth, and she was extracting them as near-complete skeletons – creatures rather than fragments.

Anning has a captivating biography, one that has frequently attracted public attention, whether in an 1865 article in Charles Dickens's magazine *All the Year Round* or the 2020 feature film, *Ammonite.*

Born in 1799 to a humble family, the daughter of a fossil-collecting cabinet-maker, as a baby Anning survived a lightning strike to a tree that killed the three women with her. She was revived by immersion in a warm bath. The lavishly garnished version of her story has it she was transformed from 'a dull and lacklustre child' into one 'lively and intelligent'.[v] Her intelligence, certainly, was beyond question. Following her first celebrated discovery, at the age of twelve, of an ichthyosaurus, the first the world had seen – sold to a local aristocrat for £23 – she went on to become a leading expert on marine reptile fossils, trusted and consulted by the respected (i.e. male) scientists of the day.

'The extraordinary thing in this young woman is that she has made herself so thoroughly acquainted with the science that the moment she finds any bones she knows to what tribe they belong,' wrote Lady Harriet Silvester, one of the visitors to Lyme who made the pilgrimage to visit the famous fossil hunter, then aged twenty-five.[vi]

Through Anning's correspondence, we learn of her undertaking dissections of cuttlefish and skate to make anatomical comparisons with fossil ink sacs and spinal columns.[vii] She was so sensitive to subtleties of proportion and difference in her specimens that she was able to tell an underdeveloped juvenile from a species variant.[viii]

Anning's life was tough. Fossil hunting was physical work, and perilous: the sea at Lyme can rise to kiss the cliffs and she had to race incoming tides to extract her treasures. The crumbling Lias brought its own hazards. One rockfall, at a promontory called Black Ven, hit so close that it killed Anning's dog Tray.[ix] Unlike the lady fossilists of her era who amassed collections as a fashionable hobby,

Anning depended on her finds for income, and the Blue Lias was capricious. Prime hunting time was winter, when storms broke up the cliffs, but a whole season could pass without the rocks yielding anything significant, leaving the shelves of her little shop bare but for quotidian ammonites and belemnites.

Born to a family of Dissenters, Anning joined the Protestant Church as an adult. Many of the great geologists of her time, the men who visited her shop, studied her fossil skeletons and consulted her, were clergymen. Yet it was a testing occupation for people of faith. The abundance of new species yielded by the Lias suggested God had created many creatures no longer evident on Earth. For creatures to have died out was felt to be a disturbing attack on the divine perfection of Creation. Nevertheless, scientists had started to discuss the idea of extinction. In 1813, the French geologist George Cuvier suggested extinctions might be caused by periodic disasters. For clergymen-geologists in the 1820 and 1830s, the Noachian deluge provided the explanation they required to reconcile faith and fossil evidence (see also Old Red Sandstone, p. 49). The sea dragons of the Blue Lias that once dominated the oceans were understood to be antediluvian: creatures from before the flood.[x]

The treasures that Anning unearthed furnished the collections of the Geological Society and Natural History Museum in London, the New York Lyceum of Natural History, the Muséum national d'Histoire naturelle in Paris, and the universities of Oxford and Cambridge.[xi] As an unmarried working-class woman, she was not permitted membership of learned societies, and authored no great papers on her discoveries. The men who studied her specimens and openly admired her expertise rose to leading positions in the new scientific field of geology, while she was left clambering over muddy rocks at Lyme Regis.

She felt the injustice keenly. In 1831 Anna Maria Pinney, one of the many young women visitors to Lyme who befriended Anning, records a conversation in which the fossil hunter is forthright: 'She says the world has used her ill and she does not care for it, according to her account these men of learning have sucked her brains, and made a great deal by publishing works, of which she had furnished the contents, while she derived none of the advantage.'[xii]

The child prodigy died young, aged forty-seven, of breast cancer, and was remembered in an exceptional address to the

Fellows of the Geological Society of London by its director, Anning's old friend Henry de la Beche. He noted that there were 'those among us in this room who know well how to appreciate the skill she employed (from the knowledge of the various works as they appeared on the subject), in developing the remains of the many fine skeletons of Ichthyosauri and Plesiosauri, which without her care would never have presented to comparative anatomists in uninjured form so desirable for their examination.'[xiii]

This was no faint praise: comparative anatomy would lead to defining revelations of the age. Among the many eminent scientists present at the meeting was the society's former secretary, Charles Darwin.

Two hundred years after Mary Anning's death, a group of geologists, palaeontologists, historians, fossil collectors, authors and others gathered in Lyme Regis for four days of talk and celebration. 'If the essence of the meeting were distilled into a single sentence it would be that Mary Anning happened to be the right person, in the right place, at the right time' concluded one present.[xiv]

Today the Blue Lias formation is a World Heritage site – part of the Jurassic Coast, site of soggy school field trips and guided fossil tours. Who can resist a surreptitious tap at freshly tumbled material in the hope of a visitation from the deep past? A casual fossil tourist doesn't stand much chance. You need to have read these cliffs like a book to know where the layers might peel apart and reveal their wonders.

CALCULI

ACCIDENTAL CONCRETIONS FORMED WITHIN THE HUMAN OR animal body, calculi, share a Latin root with calculation: *calculus* meaning 'small stone' which came to indicate the pebbles used for reckoning and computation. These concretions can occur in many locations: gallstones in the gallbladder, nephroliths in the kidneys, cystoliths in the bladder, enteroliths in the gastrointestinal tract, rhinoliths in the nasal passages, sialoliths in the salivary glands and phleboliths in the veins. Even the gunk in your navel can form an omphalolith.

If the mere thought is already too painful, best skip the display of Kathleen Lonsdale's calculi collection at the Science Museum in London. A circular, perforated case offers highlights of the thousands of specimens her team gathered around the world during the 1960s. The largest – grey as dirty chalk – are the size of fat mandarin oranges, and likewise dimpled. Some have been bisected, their innards resembling nutmegs or dates, displaying distinct layers of accreted stone like rings in a tree trunk. Most are ovoid, though some look like the rough shards of chalky limestone they basically are.

The youngest in a family of ten, born in County Kildare, Kathleen Lonsdale learned to count with yellow wooden balls rather than pebbles. This introduction to mathematics at the village school in Newbridge furnished a rare memory of her early life in Ireland. In 1908, when Lonsdale was five, her mother fled mounting unrest, and moved with her surviving offspring to Essex.[i] Lonsdale remained fascinated by calculation, and as a rare female scholarship pupil pursuing science subjects, had to travel to the Essex County High School for Boys for classes in physics, chemistry and higher mathematics. Entering university at sixteen, racking up scholarships and awards as she went, she graduated in physics from the University of London with the best grades in a decade, and entered the emerging field of X-ray crystallography.

Maths puns notwithstanding, Lonsdale's interest in calculi was an unconventional change of direction late in a career of crystalline brilliance.[ii] In a long biographical obituary written in 1974, Nobel laureate Dorothy Hodgkin wrote: 'There is a sense in which she appeared to own the whole of crystallography in her time.' One of very few female scientists to have a mineral named in their honour (Lonsdaleite is a rare form of diamond with a hexagonal lattice, found in meteorite fragments),[iii] she was, in 1945, made one of the first woman Fellows of the Royal Society. Among her many honours national and international, she was created a Dame in 1956. True to her frugal, Quaker nature, she made her own hat from 'a piece of lace, some coloured cardboard and ninepence worth of ribbons' for her investiture at Buckingham Palace.[iv]

Crystallography had evolved since the time of eighteenth-century French mineralogist Rene-Just Haüy (see Haüyne, p. 238): from the study of minerals to the atomic structure of everything. 'Most people know that diamond and sugar and salt are crystals,' Lonsdale told the BBC in 1967. 'They don't always realise that viruses and vitamins form crystals, and that metals and alloys are crystals, and so even are things like gallstones, kidney stones and bone, hair and muscle, to a certain extent, because all these things have a pattern of atoms and this affects their properties.'[v]

Lonsdale's study of calculi united two distinct sides of her personality: the physicist, and the humanitarian. A child of the First World War, her home in Essex lay beneath the flight path for Zeppelins heading towards London. Often, she did her homework by candlelight under the kitchen table during blackouts, or in the small hours of the morning after an air raid. In 1916, a Zeppelin was shot down, descending to Earth in a raging ball of flames. She remembered her mother weeping in horror, crying 'Oh, the poor men, the poor men.' 'But, mother,' her siblings said, 'they are Germans.' 'Yes, I know,' she replied, 'but they are boys.'[vi]

Sharing her mother's horror of violence, Lonsdale became an outspoken pacifist, and, after 1945, an advocate for nuclear disarmament, advising on and attending the early Pugwash Conferences on Science and World Affairs.[vii] In 1939, with three children of her own, she made a conscientious objection to civil defence duties and was sentenced to a month in Holloway prison. ('Do the police come for one,' she wondered. 'Or do I just have to go to prison by myself?') Her husband later identified her time in prison as formative, endowing her with an abiding concern for the wellbeing of others, and ease in speaking to anyone. 'Before prison it might have bothered her to go to Buckingham Palace. Afterwards, Holloway or Buckingham Palace were all the same.'[viii]

She became a prison visitor and campaigner for reform, even touring a women's detention facility in Moscow during a peace mission. In Japan, she arrived to find her hotel room crammed with flowers. A story had run in the paper that she had been incarcerated for refusing to work on the atomic bomb. Embarrassed, she asked the paper to run a correction, but clarified that she would have opted for jail under those circumstances. More flowers arrived the following day, spilling over into buckets lining the street.[ix]

In 1962 the Chief Medical Officer of the Salvation Army asked Lonsdale for help in analysing bladder stones collected while working in India as a urologist. Many had come from children living in poverty, with clusters of cases in quite narrow geographical regions: he hoped analysis would help medical researchers identify a cause and perhaps a way to prevent their formation. The unit Lonsdale established ended up studying some 3,000 samples of calculi, including historic specimens in museum collections.

Lonsdale published on the subject a few years before she died, of cancer, in 1971. Her article for *Science* opens with the thigh-clenching revelation that 'the largest human stone recorded weighed over 1.36 kilograms.' (It had occupied the bladder of Sir Walter Ogilvie of Dundee.) She goes on to describe calculi accreting layer by layer as saturated fluids leave deposits, like objects lithifying in a mineral spring. Her analysis showed different kinds of stone formations in different groups of bodies, with bladder stones among children often originating from a period of dehydration, and adult kidney stones more likely to form in those with a sedentary occupation, a particular hazard for pilots.[x] Although she died before she could see her research applied, Lonsdale's discoveries guided future treatment of the condition.[xi]

03

COPROLITE

IN 1829, GEOLOGIST HENRY DE LA BECHE DEDICATED A COMIC SKETCH to his friend William Buckland. The canon of Christ Church, Oxford, and president of the Geological Society – identified by his mortarboard, gown and geologist's hammer – appears in heroic pose at the mouth of a magnificent cavern. Arms thrown wide, Buckland surveys the scene before him: the rocky ledges are populated with deer, bear, jaguar and hyena, all captured in the act of defecation. Through the air swoop terrifying winged reptiles, unleashing a cascade of droppings as they soar past. On the banks of a lake below, fearsome crocodiles raise their tails, sending forth dung into the depths. No doubt the paddling ichthyosauri are pooping away out of sight beneath the waves. Even the cavern roof is borne aloft on twisting, turd-like columns.

De La Beche's drawing – *A Coprolitic Vision* – captured Buckland in his element, out in the field, getting his hands dirty with the remains of life forms ancient and modern. Coprolites – petrified poo – fascinated him. In 1834 he commissioned a *pietra dura* table-top for himself, inlaid with a striking collection of fish coprolites found near Edinburgh. Polished and bisected, the coprolites present a series of stony starbursts, each within a neat mahogany-coloured oval.[i] Buckland was a strange and brilliant man, laced with contradictory beliefs and sentiments, and it was often hard to tell where the joshing eccentric ended and the earnest scientist began, but his interest in coprolites was serious, and indeed provided an essential clue in the paleontological mystery that would make him a celebrity.

Buckland was Oxford University's first Reader in Geology, a position created for him with help of an endowment from George, Prince of Wales, then the Prince Regent.[ii] The letter Buckland wrote the future George IV in support of his request neatly illustrates tensions between science and religion in the period: he describes geology to be a subject 'of so much National Importance, and so liable to be perverted . . . [against] the interests of Revealed Religions'.[iii] Buckland, an ordained minister, was proffering his services not only as a scientist, but as one who would interpret his discoveries through the lens of the Anglican Church. Such blinkering would, alas, lead to some egregious errors of interpretation.

Buckland's interest in matters digestive was nose to tail: he and his son, the biologist Frank Buckland, became notorious for their zoöphagy, and an apparent commitment to eating their way through the animal kingdom, beetle to bear. In his *Praeterita*, the Victorian critic John Ruskin writes of legendary breakfasts with Buckland's large family in Oxford when he was an undergraduate, noting: 'I have always regretted a day . . . on which I missed a delicate toast of mice.'

The (perhaps apocryphal) tales of Buckland's oral fixation include an episode with a miraculous stain of holy blood in a church, which Buckland licked and pronounced to be bat urine. Another, at a dinner given by the Archbishop of York, featured the preserved remains of the heart of Louis XIV, passed around the table for inspection by guests. Buckland purportedly announced: 'I have eaten many strange things, but I have never eaten the heart of a king before' and consumed it.[iv]

His family menagerie included, at various times, a monkey, a bear and a hyena. Whether these shuffled off this mortal coil onto a chaffing dish and a bed of parsley is not recorded. He certainly had a hyena skull conveniently to hand during one of his famously animated lectures. The eminent physician, Henry Acland, a student of Buckland's, recalled him pacing with the thing in his hand before dashing up the steps and

> pointing the hyena full in my face – 'What rules the world?'
> 'Haven't an idea,' I said.
> 'The stomach, sir,' he cried (again mounting his rostrum),
> 'rules the world. The great ones eat the less, and the less,
> the lesser still.'[v]

It is little wonder that the excrement of ancient creatures would fascinate Buckland.

Billy the hyena had come into Buckland's possession to assist with an unusual scientific enquiry. In 1821 workmen in Kirkdale, Yorkshire exposed the mouth of a cave high up in the face of a rock: they ejected rubble and bones found in the mouth of the cave, assuming them to be the remains of cattle. A doctor travelling through the area noticed the bones and realised that they were something far odder: following exploration of the cave, Buckland was summoned.

He found the floor of the cave buried beneath half a metre of fine mud, over which remained, in the undisturbed portions, a crust of stalactite grown down from the ceiling and 'shooting across like ice on the surface of water, or cream on a pan of milk'.[vi] Embedded within the mud, Buckland found an enormous quantity of bones and teeth. They lay so deep that in places 'the upper ends of the bones [were] projecting like the legs of pigeons through a pie crust.'[vii] The mud had preserved the bones in remarkable condition: Buckland immersed a few in an acid bath and found that nearly all of their gelatine remained present.

With the assistance of anatomical studies made in Paris by George Cuvier, Buckland eventually identified the animal owners of the bones to have included hyena, tiger, bear, wolf, elephant, rhinoceros, hippopotamus, horse, ox, deer and numerous smaller mammal and bird species. All the bones were fragmented, with only teeth and the hardest joints left intact. Many of the remains were hyena: Buckland estimated fragments of two or three hundred to have been present in the cave.

His remarkable deduction was that the cave had for many years been a hyena den. 'This conjecture is rendered almost certain by the discovery I made, of many small balls of the solid calcareous excrement of an animal that had fed on bones.' After providing a vivid description of the shape, colour and texture of these balls, as well as noting that some contain 'undigested minute fragments of the enamel of teeth', Buckland notes that he was aided by 'the keeper of the menagerie at Exeter Change' who identified them as 'the faeces of the spotted Cape Hyaena, which he stated to be greedy of bones beyond all other beasts under his care'.[viii]

Well versed in scientific developments in continental Europe, Buckland was able to compare his discovery with similar exotic fossils recently identified at sites in France and Germany. His description and examination of the evidence in the Kirkdale cave is methodical: the problem for Buckland was then squaring the discovery of sub-Saharan fauna in a Yorkshire cave with the story of creation presented in the Bible. His discoveries were thus explained as 'inhabitants of antediluvian Yorkshire', and the book in which he presented them was titled *Reliquiae Diluvianae* – remains of the flood. Today the bones are thought to be about 75,000 years old – a span of time that would have been incomprehensible to Buckland and his contemporaries.

Buckland allowed for a great complexity of life to have existed before the flood by means of a creative interpretation of Genesis that permitted 'in the beginning' to indicate a large and unspecified quantity of time, and for many centuries to pass between the days of creation (see also: Old Red Sandstone, p. 49). Thus the fossils then starting to emerge from cliffs and quarries were explained as stages in the creation story, in which life was given form in procession from the lowliest invertebrates, through to reptiles and mammals, and the ultimate expression of God's glory: man.

Three years after his triumph at the Kirkland cave, Buckland announced something momentous at the February meeting of the Geological Society: he had identified a new species of ancient reptile, of vast size, and unlike anything seen before. He called his discovery the *Megalosaurus*. The largest femur in his possession was eighty-four centimetres, so, imagining the beast to have shared the form and proportions of a modern lizard, he estimated it to have been some twelve metres in length.[ix] In 1841, Richard Owen, an eminent expert on vertebrate anatomy, identified *Megalosaurus* as part of a distinct tribe of vast prehistoric reptiles, 'the *Dinosauria*' meaning 'fearfully great lizard'.[x]

No soft tissue remained to flesh out the forms of these great creatures, but fossilised faeces provided a clue as to their interior workings: they offered a portrait from within. In Buckland's carnivorous worldview, coprolites offered a window onto life and death on ancient Earth, onto who was eating what, or perhaps who.

#04

CORAL

NOWHERE IS THE FATE OF LIFE AND STONE MORE DELICATELY interlaced than coral. Some corals are soft; but those that build great crenelated palaces around themselves do so in limestone. Even in the cooler reaches of the northern hemisphere, we find the ancient remains of coral sometimes in slabs and blocks of building stone, evidence of warm, shallow seas that once washed this territory when it was in tropical latitudes. The fragments of coral we find are not *in* the limestone so much as they *are* the limestone, and vice versa.

The tiny coral polyp, at most three millimetres long, lives in an intense symbiotic network. Each coral is made up of hundreds of thousands of individual polyps, and in turn is part of a vast colony of diverse corals – there are about 600 species in Australia's Great Barrier Reef alone. Some corals reproduce asexually, but those that spawn coordinate the release of sperm and eggs to the same twilight moment on the same day across and between colonies. Corals are highly sensitive to environmental fluctuations. They need clear, shallow, warm water: if the sea level drops they will be exposed; if it rises they will drown; if the water is clouded with sediment, or overheats, they will die.

The coral polyp's most extraordinary relationship is with tiny photosynthetic algae – zooxanthellae – that live in their tissues. The polyp provides the algae with a protected environment and compounds required for photosynthesis. The zooxanthellae in return transfer glucose, glycerol and amino acids produced through photosynthesis back into the tissues of their coral hosts. Nutrients pumped into the coral polyp by the zooxanthellae also allow them to produce calcium carbonate and progressively regenerate and expand their limestone home.[i] The vivid colour of living coral colonies is a product of the zooxanthellae too: when coral bleaches, that's because the algae has left.

Pondering polyps' reliance on clustered social structures, and their symbiotic interspecies relationships, ecofeminist Donna Haraway notes: 'We are all corals now.'[ii] In the Great Barrier Reef, as in other mighty coral conurbations, those relationships extend above and far beyond the water. The reef is a true barrier, protecting the land from the fury of the sea. The largest living structure on Earth – 64,000 people depend on it for their livelihoods – its contribution to the environment is incalculable.[iii] For indigenous people of Queensland's Sea Country and Torres Strait Islands, the reef is both a living cultural site and a traditional resource.

At the end of the last Ice Age, around 10,000 years ago, Sea Country flooded. This ancestral memory is embedded in the creation story told by Gudju Gudju Fourmile, an elder of the Gimuy Walubara Yidinji people. It tells of a man, Gunya, fishing with his wife where the Great Barrier Reef now stands: food had become scarce and they could not find fish. Sensing movement in the water, Gunya threw his spear, hitting not a fish, but the sacred stingray Morijum. Furious, Morijum flapped his wings, and as he did so, the water rose higher and higher. Gunya and his wife paddled back to land and heated up rocks in the fire, throwing them into the ocean to calm the water and stop it rising.[iv]

Twenty thousand years ago, the sea level on the Queensland coast was 120 metres lower than it is today: land bridges connected mainland Australia with the Torres Straits Islands and Papua New Guinea. Although many stories were lost in the years indigenous languages were suppressed, among the seventy or so Traditional Owner groups whose sea country includes parts of the Great Barrier Reef,[v] references to these land bridges and now flooded territories endure. These stories date back perhaps 12,000 or 13,000 years.[vi]

Like the people, the colonies of polyps gradually shifted over thousands of years of warming at the end of the Ice Age, building new coral structures in waters of optimal depth and distance from the shore. Indeed the Great Barrier Reef has shifted backwards and forwards many times, weathering multiple cycles of global warming and cooling. The reef in its present form is young in geological terms, but grows on a platform of coral skeletons and limestone dating back millions of years.

Coral can respond to climate change, but not at the speed it is happening now: at the end of the last Ice Age, the temperature rose by 5 or 6 °C, shifting by about 1 °C every 1,000 years. Anthropogenic warming has caused an increase of 1 °C in a century. In the short term, coral is also under threat from aggressive fishing practices, from pesticides and pollutants, from excess sediment running off the land and clouding the water, and from heatwaves. High temperatures cause coral bleaching, and there have been three pan-tropical episodes since the late 1990s. Coral species can recover from bleaching, but not that fast: even the most avid colonisers will take ten to fifteen years – as global temperatures continue to rise, it is unlikely that another severe bleaching event will not happen within this time.[vii]

In 2007, Queensland started an Indigenous Land and Sea Ranger programme to draw on 'indigenous biocultural knowledge systems and Western science' for tending the reef.[viii] Aboriginal people and Torres Strait Islanders stand to lose the most from climate change, which has already brought rising sea levels, strong unpredictable winds, coastal erosion and high tides.[ix] As Traditional Owners of the Great Barrier Reef, their culture and livelihoods are rooted in its rich ecosystem. The success of the programme, which has twice been expanded, is a rare good news story in an otherwise bleak environmental picture.

After the bleaching episode in 2017, environmentalist Andy Ridley formed Citizens of the Great Barrier Reef, inviting everyone that took a boat out on the reef to document the condition of the coral. 'The Great Barrier Reef is the same size as Germany,' says Ridley. 'And one thing we realised was that 40 per cent of it had never been surveyed.'[x] With their citizen science initiative, 150 sections of reef were surveyed, by everyone from diving instructors to tugboat captains. Ridley and his team want to identify coral colonies healthy enough to spawn and repopulate bleached reefs, then ensure they are protected: they are the best hope for the future.

The Great Barrier Reef is not the beginning and end of coral's story: there are reefs and colonies throughout the oceans of the world, and they have played into the rites and belief systems of many cultures. Red and pink branching corals prized as jewels and protective amulets are from the deep waters of the Mediterranean, Japan and Taiwan. Piero della Francesca paints the infant Christ adorned in coral: thick branches of it were used to soothe teething, and to ward off danger, but here the bloodlike branches are also a prefiguring. Even into the nineteenth century the infants of wealthy families in Europe and the Americas were given coral amulets decorated with silver bells.

In the fifteenth century, Portuguese traders introduced red coral to the Kingdom of Benin. The colour was already prized for its associations with power, blood and danger, and coral became an essential part of royal regalia, used in beaded tunics, shirts, headdresses and jewellery. A French visitor to Benin in the early nineteenth century reported that the king's coral shirt weighed nine kilograms.[xi] Those wishing to adorn themselves – or their infants – in red stone branches will need to raid grandma's jewellery box. Mediterranean coral, too, is protected.

In the *Metamorphoses*, Ovid described the creation of corals as a chance event. Perseus, finding the lovely Andromeda chained to the rocks off the coast of Ethiopia, offers to destroy the sea monster menacing her in return for her hand. Beast slain, he bends to wash the blood from his hands, and makes a bed of seaweed on which to place the severed head of Medusa. The gorgon turns the swaying seaweed leaves to stone, and the nymphs try out the severed head on other plants beneath the waves, then spread coral seeds around the sea.[xii]

Medusa is a fitting emblem for coral: the mortal creature that turns flesh to stone. The gorgon's head was ubiquitous on amulets and considered a potent defence in stone fortifications.[xiii] Just as a Barrier Reef protects the coast behind it from the violence of the ocean, so Medusa's head defends the citadel walls from external threat. The tiny little coral polyp has its own defence – not only a stony citadel, but venomous tentacles that stun its tiny prey. It's a small relative of jellyfish, in French, known aptly as *la méduse*.

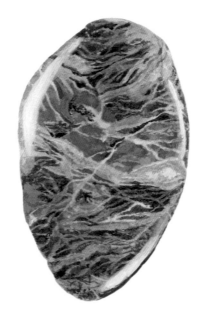

05

LEWISIAN
GNEISS

L EWISIAN GNEISS — NAMED AFTER THE ISLE OF LEWIS, BUT FOUND across the north-western extremes of Scotland – is the oldest group of rocks in Britain. Their raw material formed some 2.8 billion years ago and since then they have been heated, warped, twisted, crushed and tormented until the drama of their long lives is wrought in their fretted faces. In some areas it looks as though a calligrapher has taken an ink brush to the stony landscape and patterned it with a rhythmic abundance of strokes, some broad and banded, others rippling like light on a brook.

Most of the stony stuff that makes up the Lewisian Gneiss was igneous – granites, gabbros and the like, born in the hot belly of the Earth and cooled slowly beneath its surface, solidifying into rocks flecked with large crystals. Some of the material now bound unrecognisably into the gneiss was, mind-bendingly, sedimentary – sandstones formed from eroded debris of an even earlier generation of rocks. All have since metamorphosed: pressure-cooked at tremendous heat, as deep as forty-five kilometres inside the Earth.[i]

The Lewisian Gneiss has come a long way, moving not just up and down through hot depths of the Earth, but across its surface too. Like most of Scotland and the north of Ireland, its journey around the Earth's surface has been a different one to that of England and Wales. For a long period, 550 to 300 million years ago, Scotland and England were on different landmasses, separated by the mighty Iapetus Ocean (a proto-Atlantic named after the Titan who fathered Atlas).[ii] After a meet-cute on the supercontinent Pangaea, Scotland left North America to couple up with Europe.

The existence of an ancient supercontinent was first proposed in 1915 by German meteorologist, astronomer, geophysicist, balloonist and all-round adventure type, Alfred Wegener. Others – including the sixteenth-century philosopher Francis Bacon – had noted a jigsaw-like correspondence in the coastlines of South America and Africa. Taking the snug-fitting landmasses as his starting point, Wegener drew on his many fields of expertise to deduce how and when the landmasses of the Earth may have connected. Evidence of glacial activity in the Sahara Desert, and the presence of tropical fossils inside the arctic circle led Wegener to suggest that the Earth's great landmasses moved, a phenomenon he called continental displacement. Other evidence, such as shared fossil species found on now distant continents, suggested to him that these moving landmasses had once abutted one another.

In his book *Die Entstehung der Kontinente* (*The Origin of Continents*, 1915), Wegener proposed not only that continents shifted, but that during the Carboniferous period, all had been united in the single vast landmass that we now call Pangaea. Amid mounting controversy and debate, Wegener's book went through a number of editions – each revised – and in 1922 was published in English. What Wegener could not identify was the force propelling the motion of the continental plates. It took until 1963, after exploration of mid-ocean ridges, and the discovery that new seabed was continually being formed following the movement of the continental shelf to either side, for Wegener's conundrum to be solved and plate tectonics to be widely accepted (see: Basalt, p. 174).[iii]

The adventuresome meteorologist was not around to appreciate the vindication of his theory: he died in 1930, battling terrible conditions to carry supplies to a remote waystation during his third expedition to arctic Greenland. It was November, and Wegener and his companion Rasmus Villumsen had to ski back across the ice in punishing cold in the darkness of polar night. They never made it to the coast. 'Next spring a search party found Wegener's body almost exactly half-way down the trail, neatly buried in his sleeping bag, between his skis stuck upright in the firn. The dogs, the sled, and Villumsen were never found.'[iv]

Pangaea is now considered only the most recent supercontinent in a continuous cycle in which tectonic plates collide and separate over hundreds of millions of years. The movement is exquisitely slow – about 2.5 centimetres a year – but has been responsible for, among other things, the formation of the world's great mountain ranges. Before Pangaea came Pannotia, before that Rodinia, Nuna, Kenorland, and before that perhaps two smaller landmasses, Ur and Vaalbara. The picture becomes more and more sketchy the further you project into the past: beyond a certain point, there are no helpful fossils, or even traces of microscopic organisms, to help piece the puzzle together.

Great mountain chains were pushed up about 300 million years ago as Pangaea assembled in the southern hemisphere. As the Iapetus Ocean closed, and the ancient continent of Laurentia (territory of much of present-day North America) collided with the plates of Gondwana, Baltica and Avalonia, mountains as great as the Himalayas formed along its margin. When Pangaea started to break apart, about 175 million years ago, this mountain range split with parts of the ancient continent of Laurentia remaining attached to modern-day Europe. On one side of the Atlantic, the eroded remains of these mountains endure as the Appalachians, on the other they stretch from the north of Ireland and north-west of Scotland all the way to the Svalbard archipelago in Norway (see also: Coal, p. 233).[v]

Lewisian Gneiss, once part of the same landmass as Canada, ancient souvenir of deep-Earth drama and veteran of many supercontinent cycles, seems imbued with life. Like the twisting anthropomorphic trunks of olive trees that inspired the ancient poets of Southern Europe to spin tales of metamorphosis, this rock appears caught in movement. On Lewis, vibrant slabs and spires of gneiss form prehistoric structures, most stirringly at Calanais:[vi] great avenues intersecting at a circle girding a tall central pillar. Scholar of the megalithic Aubrey Burl lists the various functions ascribed to Calanais over the centuries, with theories ranging from it being the remains of a Gothic court to 'a landing-mark for Martians who also happened to be Beaker Folk'.[vii]

Burl finds the most compelling explanation for the stones at Calanais in the peculiarities of their geography. Lewis sits fifty-eight degrees north, a latitude at which a huge 'major moon' appears at a position so low it seems to touch the Earth every 18.6 years. This is a complex cycle, and charting the interval at which these extreme lunar spectacles occurred would have been the work of many generations. Improbable as that may sound, Burl then cites a Greek account from the first century BCE that appears to give an account of the lunar sightline in operation, over 2,000 years after the stones were erected. This early voyager described an island in the north of Britain that carried a 'spherical' lunar temple. 'The moon as viewed from this island appears to be but a little distance from the earth.' The god of this temple is noted as visiting the island 'every nineteen years'.[viii]

Many thousands of years after they were erected, the stones at Calanais acquired new legends. They were known as *Fir Bhreige* – false men – and were said to be the petrified remains of heathen giants, punished by the fifth-century Irish saint St Kieran for failing to succumb to his preaching.[ix] Lewisian Gneiss has no need for such fanciful mythologising: it's been through adventures and metamorphoses enough.

#06

PEARL

O N REFLECTION, THERE IS LITTLE ABOUT A PEARL TO INSPIRE thoughts of purity. For oysters, mussels and other bivalves, a natural pearl is a defence mechanism, a nacre coating for an irritant – often a parasite – that has made its way into the fleshy interior. All that lustrous shimmer is merely the decorative death-bed of an invasive worm. To create a cultured pearl, according to the process patented by Japanese pearl magnate Kōkichi Mikimoto in 1916, the host oyster is forced open, a slit is cut in its gonad and a two-millimetre square from the mantle tissue of a donor oyster inserted with a little shell bead. Whatever a pearl's start in life, things do not go well for the oyster.

Parasites and gonad piercing notwithstanding, pearls' other-worldly sheen and rarity have brought them associations with the noble and divine across centuries and cultures. In the third-century Syriac poem known as 'Hymn of the Pearl', a beautiful young nobleman has his jewelled robes and tailored gown stripped from him and is sent west to Egypt to fetch 'the one Pearl . . ./ that lies in the Sea,/ Hard by the loud-breathing Serpent.'

Taken into slavery by the 'unclean' Egyptians, he is roused by a magic missive from his parents that flies to him like a bird, returning him to his quest. Invigorated, he enchants the serpent and returns triumphant with the pearl, earning an audience at the 'Court/ Of the King of Kings'.[i]

While this allegorical pearl is an emblem for the pursuit of holy knowledge, a true quest for a great pearl in Egyptian waters would have been wishful thinking. The Red Sea in the third century yielded inferior stock, not up to the standards of the Roman elite, who prized pearl above all other jewels, and were greedy for magnificent specimens. Egypt was indeed where Romans bought their pearls, but they had arrived there from the Gulf of Persia, or the waters off Sri Lanka.

The genesis of pearls was long a mystery: the sixth-century scholar Isidore of Seville suggested that oysters would seek the shore at night and 'conceive the pearls by means of celestial dew'.[ii]

He notes that 'Britannia' (Britain), 'Taprobane' (Sri Lanka) and India are rich in pearls. Those from 'Britannia' came from fresh water mussels, once abundant, since over-fished, now protected.

By the sixteenth century, the 'celestial dew' impregnating oyster with pearl was accepted as a metaphor for the Virgin Mary's miraculous conception. As a nuptial gift, Prince Philip of Spain gave the godly English Queen Mary a diamond brooch from which dangled an enormous pear-shaped pearl. At once pure and magnificent, the vast jewel, known as the Mary Tudor Pearl, appears in portraits of the queen, hanging from a square pendant at her chest.[iii] The jewel was apparently much coveted by her younger sister Elizabeth, but Mary insisted all jewellery given to her by Philip be returned to him on her death.[iv]

Elizabeth I, on taking the throne, gave her privateers licence to ransack Spanish ships returning from the Americas, and deliver their pearls and other jewels to her (see also: Ruby, p. 54). In her portraits the Virgin Queen appears bristling with pearls: they nestle like buds of snow in her hair, garland her waist and shoulders, and shimmer from the embroidered surfaces of her gowns: an emblem of chastity, they were part of her cult and her diplomatic image. In 1559, the Venetian ambassador Il Schifanoya reported that, meeting with ambassadors from France, the queen was 'dressed entirely in purple velvet, with so much gold and so many pearls and jewels that it added much to her beauty.'[v]

The men and women of the court covered themselves with pearls in homage. Henry Percy, newly made earl in 1585, spent over £1,000 on jewellery, including 440 pearls to decorate his clothes. Such lavish garb was the preserve of favoured nobility: among the many sumptuary laws forbidding excesses of dress during Elizabeth's reign is specific interdiction of anyone below the rank of baroness wearing cloth or silk mixed with pearls.[vi]

Kōkichi Mikimoto was not the first to experiment with cultured pearls – since the fifth century in China, lead medallions of the Buddha have been placed inside fresh-water mussels to be coated in mother of pearl. The Chinese had long been able to create hemispheric *mabe* pearls using the same process.[vii] Mikimoto was the first to culture full pearls successfully, and in doing so made the stones an accessible luxury. By 1930, the cultured spheres rolling out across the world caused the price of all pearls to crash.

These pearls were not fakes: they had a mother of pearl bead at their core and their layers of nacre were built by irritated oysters.

In the same decade Mikimoto's perfect spheres first dominated the world market, a rather stranger pearl created a sensation in the US. Told in the breathless imperial style of a boy's magazine adventure – replete with 'natives' agog at white saviour medicine – Wilburn Dowell Cobb's tale of how he came to possess the Pearl of Allah was published by *Natural History* magazine. The story ran as the pearl, which resembled a wad of chewed gum a little larger than a human brain, went on display at Ripley's Believe It or Not! odditorium in New York in 1939.

Set in 'the outer reaches of the Philippine Islands', Cobb's yarn opens with tragedy – a diver drowned with his foot caught in a giant Tridacna clam. Once the unfortunate soul was freed, the clam revealed its treasure – a vast pearl, over six kilograms in weight, in the wrinkled surface of which the local 'chief' Panglima Pisi sees a likeness of the prophet Mohammed. Pisi tells Cobb that he will not part with it at any price. Many months later, Cobb administers the malaria remedy Atabrine to Pisi's son, and nurses him back to vigour: as payment, he is given the pearl.[viii]

Like a bead acquiring nacre, the Pearl of Allah accumulated layer after layer of myth and mystery. While on display at Ripley's on Broadway, the pearl was spotted by a mysterious Mr Lee, who pronounced it the long-lost pearl of philosopher Lao Tzu: a 2,400-year-old amulet carved with the faces of Buddha, Confucius, and Lao Tzu himself, which had been placed in an oyster to acquire glossy nacre. Lee offered half a million dollars, hinting that the pearl was worth seven times as much. Cobb, like Pisi before him, refused to sell.[ix]

Travelling the world with his pearl, Cobb kept building the myth. At his death in 1979, his family invited the Internal Revenue Service to make a more conservative estimate. They agreed on $200,000, for which the pearl was sold to a medallion-wearing chancer called Victor Barbish. Barbish polished Cobb's story and then some: the near-priceless stone was always on the point of being sold when some picturesque calamity hit. Inevitably Barbish just needed a little injection of cash – perhaps in return for shares in the ever-inflating sale price – to tide him over. His stories were wild – at one point he claimed Osama Bin Laden had attempted to buy the pearl as a gift for Saddam Hussein – yet believed. After all: the pearl was real, *The Guinness Book of Records* listed it as the world's largest, and Barbish owned it.

From 1979 until his death in 2008, Barbish used the pearl to leverage ever more outrageous cons. Willing 'experts' helped fluff the pearl's age back to 600 BCE – around the time of Lao Tzu – and inflated its value as far north as $53 million.

There are pearls, and there are pearls: this was an ugly concretion, not a glossy jewel. As of 2016, the 'Pearl of Allah' was no longer even the world's largest. Invited to estimate its value recently, the British auction house Bonham's suggested $100,000: half the price Barbish had paid when it was sold cheap as a tax liability in 1979. Since Barbish's death, the pearl has been caught tight in the jaws of the legal system, involved in disputes over shared ownership, unrecovered debts and even a murder case. Where once it was claimed the pearl was too precious to sell at any price, now it is of value as a curiosity, with a very real history of showmen, con artists and other parasites trailing behind it.

#07

SLATE

Slate is the stone of both scholarship and delinquency – the blackboard and the billiard table – and decks the roofs of schoolrooms and pool halls alike.[i] A hard, dark, rock that offers a smooth, regular surface, the precision by which it can be split into thin sheets is described by the revolting term slaty cleavage (surely first choice for a drag name in a geological cabaret?).

In Wales, heart of the British slate industry, the phenomenon of the rock splitting into sheets is called *hollt*.[ii] Slate has been both mined and quarried in North Wales, leaving huge scars on the ancient dappled mountains, and deep tunnels within them. A beautiful study of the Eryri (Snowdonia) national park, published just after the Second World War describes the slate mines around Blaenau Ffestiniog as 'the largest of their kind in the world' with chambers almost 100 metres long and thirty metres high: 'in one mine several such chambers are or have been worked on as many as twenty-four levels, the lowest of which is more than 1,500 feet [457 m] from the top of the incline that gives access to it.'[iii]

Unknown to the authors, earlier that decade the mines around Blaenau Ffestiniog had provided an unlikely home for paintings by Rubens, Titian, Van Dyke and Velázquez. In the late 1930s, with war looming, Kenneth Clark, director of the National Gallery, feared the central London institution and its treasures were vulnerable to attack. The Gallery, as a large edifice flanking Trafalgar Square, was a particularly eye-catching landmark from the air. Ten days before the declaration of war on 3 September 1939, the paintings were removed and temporarily deposited at castles and libraries in Wales, but Clark needed to find a secure location. It was suggested that the paintings be shipped to Canada, but Winston Churchill objected: 'Hide them in caves and cellars, but not one picture shall leave this island.'[iv]

By the summer of 1941, all the 2,000 paintings were accommodated in the depths of a modified slate mine. Clark's fears were justified: the National Gallery sustained serious bomb damage, hit nine times between October 1940 and April 1941. When the building was reconstructed, the conservators drew on their observation of temperature fluctuations and humidity in the mine to include such novelties as air conditioning in the Gallery itself.

The construction boom that accompanied the Industrial Revolution had generated a huge appetite for slate, used as roofing for factories and for the snaking terraces of houses that accommodated Britain's rapidly urbanising population. By the end of the nineteenth century, Wales was home to 80 per cent of the British slate industry, with 17,000 men mining and processing 485,000 tons in a year.[v]

Slate has been quarried in Wales since Roman times, used in the construction of Segontium, a fort overlooking the Menai Strait, defending the coast of North Wales against raids from Ireland. Wales' mineral wealth made it valuable territory: the Romans maintained Segontium as an administrative and military base for over three centuries, from 77 to about 394 CE. They called the country Cambria – a name that lives on in the geologic timescale as the Cambrian Period, 541 to 485 million years ago. The entire four billion years before that can be lumped together as the Precambrian supereon.

Wales was the hammering ground of competitive early geologists, who tumbled over one another to identify formations.[vi] Around the world, the period 485 million to 443 million years ago is known as the Ordovician, named after the Celtic Ordovices tribe of North Wales, and 443 million to 419 million years ago is Silurian for the Silures of the South. A patriotic Welsh geologist might justifiably consider over 90 per cent of Earth's geological time to have been named in her country's honour. (There is of course no temporal correlation between the tribes and the geological periods: the Ordovices and Silures were active at the turn of the first century CE, and were around to put up resistance to the invading Romans.)

Slate is metamorphosed mudstone: the fine blue-grey stone mined from the old slopes of Eryri was once a gloop of fine silt eroded from the mountains that towered over the ancient microcontinent of Avalonia. At the start of the Cambrian, the silt built up in deep oceans that overlay the continental crust: hundreds of millions of years later, deep underground, it was metamorphosed by the heat and pressure of a mountain-building episode that accompanied Avalonia's various slow collisions as the supercontinent Pangaea formed. Slate splits cleanly because the mica within it aligns during this metamorphosis: it is not splitting along the layers in which it formed.

In his book *The Planet in a Pebble*, British geologist Jan Zalasiewicz reads the history of the Earth through a piece of slate found on a Welsh beach. It's one of those flattish, roundish pebbles crisscrossed with white quartz: irresistible for skimming through surf, or slipping into your coat pocket, but otherwise quite ordinary. What is remarkable in Zalasiewicz's account is the degree to which the pebble is produced by, transformed by and, for hundreds of thousands of years, actually *is* living matter.

Together with acid rain, the very first land-based algae and fungi helped break down the rocky surfaces of Avalonia into light clay particles, which were washed into the sea and slowly settled. The muddy mineral sediment on the sea floor is nutritious, described by Zalasiewicz as 'a sedimentary broth', eaten, digested and excreted by an army of worms and burrowing molluscs.[vii] Passage through the guts of these creatures alters the chemistry of the silt. Thin wisps of black carbon in the slate document 'a continuous procession of death' – plankton corpses that fell and settled on the sea floor together with an endless glitter shower of tumbling faecal pellets.

Even in anoxic conditions, the depths of the sea would have been home to complex colonies of microbes. Embedded in the stone are the shelly fossil remains of tiny life forms, including saw-tooth-like graptolites. Even after 100,000 years, buried now beneath 100 metres of mud and sand, the stiffening stuff of our future slate is still alive with microbes.[viii] Life returns as soon as the slate is exposed to air and water through cracks in the cliff: microbes first and, later after it surfaces, lichens.

As well as slate, coal and metals, the mines of Wales have supplied writers from Shakespeare to Tolkien with a rich tradition of folk tales and fairy stories. The coblynau – antecedents to the goblins of modern fiction – were said to be about half a metre in height and 'very ugly to look upon, but extremely good natured, and warm friends of the miner'.[ix] The activities of coblynau explained the noises and apparitions of the dark, damp world underground, and the uncanny sense that the miners shared this space with forms of life invisible to the human eye – which of course they did.

#08

SULPHUR

PLATE 53 OF *CAMPI PHLEGRAEI* COULD HAVE BEEN PAINTED IN A domestic larder. At the heart of the picture is what appears to be a majestic cauliflower, stripped to its bulbous white protuberances. Before it are what look like a slab of raw beef and mound of butter. The forms to either side are perhaps joints of ham, covered in golden layers of fat. In fact all are mineral specimens – 'Volcanic rock from Solfatara' – and the oleaginous lumps and layers, accretions of sulphur. This *nature mort* is among illustrations by Peter Fabris commissioned by the eighteenth-century British diplomat, collector, connoisseur and volcano lover William Hamilton for his books on the eruptive phenomena of Naples and Sicily.[i]

In another plate, Fabris shows the elegant figure of Sir William himself inspecting the Hadean landscape of Solfatara – literally, 'sulphur place' – near Naples. At the centre of a denuded valley a benighted pit exhales creamy fumes, beneath which are mounds of the vivid yellow stone. Describing the scene, Sir William speculates that the 'hot vapour' that 'issues with violence' comes from a boiling subterranean lake heated by 'fires still deeper'. Solfatara is the crater of an ancient volcano, which last erupted in 1198. The sulphur and fumes suggest to Sir William that it is not done yet.[ii]

The *Campi Phlegraei* (Campi Flegrei in Italian) are the 'flaming fields' – a district of rumbling volcanic activity near Naples. Here the stony earth is anything but cold and dead – you can smell the feculent air emitting from its bowels, and see the living rocks generated from the depths.

Sulphur is the yellow stone of Yellowstone, considered a benighted, bubbling, hellish place by European trappers in the nineteenth century. As brimstone, it perfumes the horror of medieval hell: in the Book of Revelation, 'the fearful, and unbelieving, and the abominable, and murderers, and whoremongers, and sorcerers, and idolaters, and all liars, shall have their part in the lake which burneth with fire and brimstone: which is the second death.'[iii]

Sulphur's liveliness earned it a central role in the centuries-long quest for the philosopher's stone: a powdery substance endowed with transformative, elevating powers over man and metals. Thirteenth-century German scholar Albertus Magnus described sulphur as 'the fatness of the earth', a term that not only encompasses the buttery, lipid qualities on display in *Campi Phlegraei*, but also suggests ripeness and fertility. Of the alchemical art, he explains it is possible 'to bring about a new body, since all species of metals are produced in the earth from a commixture of sulphur and quicksilver.'[iv]

Such ideas were rooted in Plato's principle of the four elements (air, fire, earth, water) that dictate the form of matter, and Aristotle's description of rocks and metals forming from exhalations within the earth. Nature worked towards perfection, according to Aristotle, thus the finest materials were evidence of a long ripening in optimal conditions.[v] Later alchemists believed that metals could be artificially ripened and refined to achieve their ultimate form – gold.

Neither Plato nor Aristotle wrote from within the sphere of alchemy. The roots of the European tradition instead lie in gnomic aphorisms of the mysterious Hermes Trismegistus, inhabitant (if he lived) of Alexandria in the early centuries of the Roman Empire. The legacy of Hermes' obscure prose endures in the term 'hermetic': his vexing *Tabula smaragdina* (Emerald Table) is the alchemist's foundational text. It describes 'the miracles of one thing', 'the father thereof is the Sun, the mother the Moon' which will be 'the father of all works of wonder throughout the whole world', including the power to separate 'the subtle from the gross'. Hermes offered no recipes, no materials, no methodology, but sufficient ambiguity to occupy philosophers for millennia.

Sulphur and mercury were introduced to alchemical lore during the Islamic Golden Age,[vi] through the writing of Abu Musa Jabir Ibn Hayyan and Avicenna (see also: Cinnabar, p. 78). All metals were the product of these two substances, and variety was the result of different purities. The most subtle mercury and superior white sulphur produced gold; good mercury but combustible sulphur gave copper; corrupt mercury and impure sulphur, iron; and so on.[vii] 'Khem' is Egypt, birthplace of the mysterious art known in the Islamic world as *al Khem*, which passed into medieval Europe as 'alchemy'.[viii]

Thus commenced centuries of distillation, sublimation, the

application of fire and urine, burials of sealed flasks in steaming dung heaps, and instructional texts that delivered more puzzlement than gold. Alchemy has its technical roots in the Egyptian arts of metallurgy, and the production of dye and perfume.[ix] It is the origin of much scientific kit: the double saucepan known as a bain-marie was named after the alchemist Maria the Jewess, thought to have lived in Alexandria in the first century CE. What little is known of Maria comes from admiring accounts centuries after her death: she is said to have founded an alchemy academy and knew how to derive gold from plants.

There were few female alchemists, but another of this early era, known as Cleopatra (not the queen), is credited with inventing the alembic, used for distillation. Cleopatra's ideas endure only through secondary accounts. A wise woman of Alexandria, she is cited in a number of mystifying conversations on alchemy. Later practitioners largely remembered Cleopatra for alopecia remedies, still in use in medieval Europe.

It would be misleading to imagine alchemy simply as chemistry in its infancy: this was a spiritual art and the powers sought in the philosopher's stone were a means of transformation of the self as much as they were a means of material transformation. The alchemists' world was one in which all matter was animated and possessed a living soul,[x] and the language used to describe the comingling of materials was overwhelmingly that of sexual reproduction.[xi] In his *Radix Mundi* in the thirteenth century, philosopher Roger Bacon wrote of marrying the 'White Woman [mercury] to the Red Man [sulphur]', by which the 'Semen Solare [sulphur]' would impregnate the 'Matrix of Mercury, by Copulation or Conjunction' to be made one.[xii] In geology, minerals are still described as occurring within a rock 'matrix' as though they were the product of an alchemical birth.

The confounding language of alchemical texts was exacerbated by an increasingly complex symbolic vocabulary. Philosophical sulphur and mercury became symbolic entities, distinct from those elements as manifested in their natural form. During the flourishing of alchemical activity in the Renaissance, Christian symbolism was also brought into the mix. For the European alchemist, sulphur carried with it the whiff of hellfire. In mastering sulphur to produce gold, psychiatrist (and alchemy scholar) Carl Jung noted, 'it would have been natural for the alchemist to suppose that they had lured the devil out of the darkness of matter.'[xiii]

The discipline's last true adherents include Isaac Newton at the turn of the seventeenth century (see Cairngorm, p. 73). For many years, historians of science regarded Newton's extensive writings on alchemy as an embarrassing lapse, best forgotten. The grand questions of immortality, and the transformation of matter, which occupied great minds for thousands of years, had to be set aside for 'rational' science to flourish. By the late eighteenth century Sir William Hamilton and his peers flattered themselves that they lived in an age of explainable 'marvels' rather than miracles.

Among the essential divisions laid down by men of the Enlightenment were separations between humanity and nature, between matter and spirit, and between science and the uncanny. The fashion for collecting, recording, categorising and naming, exemplified by Sir William and his sulphurous rocks, was a rational riposte to the mystic superstition of alchemy. Nevertheless, for those whose pulse quickens to tremors and eruptions – volcano lovers – it is hard to banish thoughts of wickedness from the stench of sulphur.

#09

THE
UJARAALUK
UNIT

THE HADEAN IS THE TIME OF EARLY EARTH, THE 500 MILLION years following its accretion 4.567 billion years ago. Trust that doomy name: the Hadean would not have been a fun time to hang out. For the first few million years, Earth rotated on its axis every five hours, orbiting the sun in a year of 1,750 short days. The forming moon would have looked huge, only 25,000 kilometres away with red magma glowing through cracks in its crust, orbiting Earth every eighty-four hours. The proximity of the two bodies would have caused violent tidal waves of magma, the molten rocky surface cresting in mountainous bulges that rippled around both Earth and moon.[i]

The temperature above this roiling surface would have been like a blast furnace: over 5,500 °C immediately after the impact of Theia (see Moon Rock, p. 146), with an atmosphere of silicate vapour that cooled into a shower of magma droplets. The first crystals formed when Earth's surface cooled to about 1,650 °C and the planet acquired a black skin of basalt.[ii] As an ill-advised tourist visiting Earth 4.5 billion years ago, you would have been bombarded by meteors and showered by volcanic magma, all within a rich sulphurous stink: vapourised before you had time to be asphyxiated from the lack of atmospheric oxygen or sack your time-travel agent.

The Hadean Eon was defined by its lack of rock record: the basalt crust produced in this period was recycled back into the mantle. Earth's first 500 million years have been a dark era, even for the fill-in-the-gaps science of geology. Zircon crystals about four billion years old have been found in the Jack Hills conglomerate in Australia, but their original host rock has long since been destroyed. For a while the Acasta Gneiss formation in Canada's Northwest Territories contained the earliest known fragments of terrestrial crust.

That may yet prove to be the case, except that, in 2011, study of the Nuvvuagittuq Greenstone Belt in northern Quebec suggested some of its rock was as much as 4.3 billion years old. Of Hadean age, in other words, which would make it the earliest preserved crustal rock on Earth – the opening sentence in the book of Earth time. There is no neat pigeonholing this rock. The first name geologists came up with for the ultra-mashed and mixed-up magnesium and iron-rich gneiss was the 'faux-amphibolite'. Faux-anything sounded frankly insulting for rocks that have survived 'multiple phases of deformation and high-grade metamorphism', so this ancient of days was redubbed the Ujaraaluk unit.[iii]

There are patches of very ancient crustal rock – as much as 3.9 billion years old – flecked across the surface of Earth, like birthmarks, in south-west Greenland, northern Labrador, eastern Antarctica and central China.[iv] These are the rocks that could shine light into Earth's dark age, tell us what chemical changes took place, how the earliest continental crusts were formed, when and why plate tectonics started. To get there, geologists must first untangle everything that has happened to them over the last four billion years.

How do we fragile, fleshy beings grasp such timescales? Human time is written in tens and hundreds of years rather than millions. Philosopher Timothy Morton suggests we are more able to grasp an abstract notion such as 'infinity' than acclimatise to truly enormous timescales. 'There is a real sense in which it is far easier to conceive of "forever" than a very large finitude,' writes Morton. 'Forever makes you feel important. One hundred thousand years makes you wonder whether you can imagine one hundred thousand anything.'[v]

The writer John McPhee refers to this realm of vast spans – millions and billions of years – as 'deep time'. He suggests geologists are functionally bilingual, able to adapt their minds to the requirements of both human and planetary timescale: 'Geologists, dealing always with deep time, find that it seeps into their beings and affects them in various ways. They see the unbelievable swiftness with which one evolving species on the earth has learned to reach into the dirt of some tropical island and fling 747s into the sky.'[vi]

Whether or not the Ujaraaluk unit is truly Hadean or 'merely' Eoarchean (4–3.6 billion years old), these rocks have accompanied Earth through its extraordinary transformations: the origin of microbial life, over 3.5 billion years ago; the start of algal photosynthesis precipitating the Great Oxygenation Event that transformed the planet's atmosphere about 2.2 billion years ago; the building and breakup of supercontinents; a global ice age 740 million years ago;

the explosion of animal life in the oceans 530 million years ago, and, 55 million years later, the first land plants.

It has witnessed five mass extinction events, including the catastrophic Great Dying 251 million years ago, which wiped out 70 per cent of land species and 96 per cent of those in the sea. There have been dramatic cycles of global cooling and warming, long before humans came into the picture.

Thinking in such vast expanses makes Zen poet-philosophers of geologists: 'To understand Earth, you must divorce yourself from the inconsequential temporal or spatial scale of human life,' writes Robert Hazen.[vii] Perhaps the most important thing we can learn from the Ujaraaluk unit is a sense of perspective.

A LEXICON OF LITHIC LINGO

ASTERISM

Inclusions in a gemstone creating the appearance of an inner star.

BÆTYL

Or bet el – 'house of god' – a stone considered home to a divine presence.

CALCULUS

'Small stone', the pebbles used for reckoning and computation.

CARATS

Unit of weight used for diamonds and other gemstones. One carat is 200 milligrams. Not to be confused with karat, a measurement of gold purity.

CHATOYANCY

Cat's-eye streak in a gemstone.

CLEAVAGE

Tendency of certain crystalline materials to split along a defined plane, leaving a smooth surface.

CRATON

Ancient stable portion of a continent, composed of crust and upper mantle, the oldest rock on Earth.

CRYPTOCRYSTALLINE

Material composed of microscopic crystals.

DIRECT CARVING

Sculpture carved directly into stone or wood, deriving form from the material, rather than being carved according to a pre-existing design in plaster or clay.

ECOSEXUAL

Environmental activist who chooses to treat the Earth as a lover rather than a mother.

ERRATICS

Also known as glacial erratics. Boulders deposited by a glacier, often of a distinctly 'foreign' rock variety to the site they are deposited on. From Latin *errare*, 'to wander'.

FANTASY CUT

Sculptural, non-traditional gemstone cutting style pioneered by Bernd Munsteiner.

GEODE

Rounded rock in which minerals have built up and crystallised around a central hollow. In a classic formation, outer layers of chalcedony surround an inner layer of crystals.

GEOGLYPH

Design or motif carved or constructed with the durable elements of a landscape.

GLYPTIC

The art of carving or engraving, particularly used in relation to gemstones.

GRINDSTONE

A turning wheel for metalwork.

INCLUSION

Irregularities or foreign matter trapped inside a mineral during its formation, including other crystals, liquid and gas bubbles.

INTRUSION

Magma from beneath the Earth's surface that pushes up into cracks and hollows in the existing rock, solidifying before it reaches the surface.

IGNEOUS

Stone formed from cooled lava or magma. From the Latin, *igneus*, 'fiery, ardent, burning'.

KNAPPING

The shaping of flints, obsidian or chalcedony by knocking off angled flakes with a hammer stone or bone.

LAND ART

Site-specific artworks made directly in the landscape, typically using materials found on site. Also known as 'Earth Art'. Can be permanent – such as Michael Heizer's *Double Negative* (1969–70) – or ephemeral – such as Richard Long's lines formed by walking.

LAPIDARY

The art of cutting and shaping stones; a person who practises this art; a volume listing the qualities of stones and minerals.

LITHIC

Of, or relating to, stone.

LITHIFICATION

The slow process by which deposited sediment turns into rock.

LITHOPHONE

Stone emitting a bell or drum-like sound when struck.

LITHOSPHERE

Earth's crust and outermost portion of the upper mantle: the comparatively rigid, rocky shell of the planet, typically 100 km thick, which lies over a layer of hotter, partially molten rock within the upper mantle known as the asthenosphere. From greek, *lithos*, 'rock'.

MACROCRYSTALLINE

Occurring in crystals large enough to be seen with the naked eye or a simple lens.

MEGALITH

Large, usually undressed (i.e. rough) stone used in Neolithic or early Bronze Age monuments.

MENHIR

A positioned standing stone, either solo or in groups.

METAMORPHIC

Stone transformed by heat and pressure.

MOHS SCALE

Ranging from talc (1) to diamond (10), this is the scale along which stones are judged for hardness. Rather than a scale of even increments, it is ordered with the stone at each point hard enough to make a mark on the one beneath it on the scale: diamond can scratch corundum (9) which can scratch topaz (8) which in turn can scratch quartz (7) and so on, down to gypsum (2) and talc.

NACRE

Mother of pearl.

NEOLITHIC

New stone age.

NEPTUNISM

Eighteenth-century theory according to which all the stones of Earth built up at the bottom of a vast primordial ocean.

NUÉE ARDENTE

From the French, 'burning cloud', a pyroclastic flow that glows red.

OBJET DE VERTU

Luxury curio. The term translates from French as 'object of virtue', but is a British coinage for an object valued for its material, craftsmanship or antiquity.

PALEOLITHIC

Old stone age.

PEGMATITE

A class of igneous rocks composed of exceptionally large crystals.

PETROPHILIA

Love of stones.

PETROMANIA

Obsessive love of stones.

PHILOSOPHER'S STONE

Mythic alchemical substance believed to grant eternal life and transform base metals into gold.

PIEZOELECTRICITY

Electrical charge that can accumulate in crystals under stress.

PLAY OF COLOUR

Optical phenomenon in which light reveals dancing, flashing colour within stone such as opal or labradorite.

PYROCLASTIC FLOW

From the Greek *pyro*, 'fire', and *klastos*, 'broken': burning hot clouds of pumice, gas and ash cascading from a volcano during an explosive eruption.

REGOLITH

Loose dust and rock that sits above solid bedrock.

RIFT

Zone in which the lithosphere is pulling apart due to the movement of tectonic plates.

SHOWSTONE

Crystal ball used for divination.

SCRYING

Receiving visions through a medium such as crystal or obsidian.

SUPERCONTINENT

Ancient landmasses formed during episodes when continental plates have collided: Pangaea, Pannotia, Rodinia (from the Russian for motherland or birthplace, ро́дина), Nuna and Kenorland. Before that there were perhaps two smaller landmasses, dubbed Ur and Vaalbara.

TECTONIC PLATE

A section of the lithosphere. Driven by the energy of heat from the mantle rock below, tectonic plates move relative to one another, at a speed measurable in cm per year. Interaction at the plate boundaries can be responsible for earthquakes, volcanic activity and mountain-building episodes.

TEKTITE

From the Greek *tēktos*, 'molten': natural glass found around meteorite impact sites.

TOUCHSTONE

A dark and fine-grained stone, such as slate, used for assaying precious metal alloys (typically gold) according to the streak left on the surface.

TREATMENT

Colouring, bleaching, impregnating, heating, irradiating a gemstone to boost its tone, clarity and price.

TRIBOLUMINESCENCE

Producing flashes of light under stress.

UNCONFORMITY

A gap in the sedimentary rock record, often of many millions of years, for example, Powell's Unconformity in the Grand Canyon which represents a hiatus of over a billion years.

UNIFORMITARIANISM

Theory first proposed by James Hutton in the late eighteenth century, suggesting that the cycle of rock formation, erosion, sedimentation and lithification was continuous, occurring in the deep past as it does in the present day.

XENOLITH

Literally 'alien rock.' A mineral pushed up from deep in the Earth's mantle.

ENDNOTES

STONES
AND POWER

i   Octavia Rooks, 'Scientific Discovery, Imperialism, and the Geological Map', *Bluesci: Cambridge University Science Magazine*, published online 13 January 2022.

ALUNITE

i   Kermes was a red dye derived from insects. In modern Turkish *kırmızı* is the colour red.

ii   Jane Schneider, 'Peacocks and Penguins: The Political Economy of European Cloth and Colors', *American Ethnologist*, August 1978, vol. 5, no. 3, Political Economy, p. 416.

iii   Present day Izmir

iv   Quoted in Roger Osborne, *The Floating Egg* (Pimlico, 1999), p. 10.

v   Quoted in Rhys Jenkins 'The International Struggle for Manufactures as Illustrated by the History of the Alum Trade', *Science Progress in the Twentieth Century* (1906–16), vol. 9, no. 35 (John Murray, January 1915), p. 49.

vi   Charlotte Moy, 'Economic History of The Medici Family', *Economic Historian*, 8 December 2020, economic-historian.com

vii   Paula Hohti, 'When Black Became the Colour of Fashion', *Refashioning the Renaissance*, 14 February 2020, refashioningrenaissance.eu

viii   Ibid., Paula Hohti writes:'Post-mortem inventories from Venice, Florence and Siena drawn up between 1550 and 1650, demonstrate that over 40 % of all artisans' clothes that were identified by colour in the inventories were described as black.'

ix   Jenkins, 'International Struggle', p. 493.

x   Quoted in Schneider 'Peacocks and Penguins', p. 428.

xi   Osborne, *Floating Egg*, p. 15.

xii   Ibid., p. 20.

AMBER

i   Theophrastus, *On Stones*, 28–9, trans. and ed. Earle R. Caley and John F. C. Richards (Ohio State University, 1956).

ii   William Gilbert, *De Magnete*, Book II. Ch. 2. (1600).

iii   Old Prussia occupied parts of present-day Poland, Lithuania and the Kaliningrad Oblast.

iv   Paul A. Zahl 'Natural History: In an Amber Mood', *American Scholar*, spring 1978, vol. 47, no. 2, pp. 237–44 (Phi Beta Kappa Society, 1978), p. 238.

v   Ibid.

vi   Victoria Finlay, *Jewels: A Secret History* (Hodder & Stoughton, 2005), pp. 44–5.

vii   Frederick William I, King of Prussia, British Museum collection online.

viii   Finlay, *Jewels*, p. 45.

ix   Quoted in Zahl, 'Natural History', p. 237.

x   By Diogenes Laertius in his discussion of Thales, quoted in the commentary to Theophrastus *On Stones:* 'Aristotle and Hippias say that, judging by the behaviour of the lodestone and amber, he also attributed souls to lifeless things.'

BLACK SHALE

i   Joe Hoffman, 'Potential Health and Environmental Effects of Hydrofracking in the Williston Basin, Montana', published online by Teach the Earth, 2012)

ii   Eric Konigsberg, 'Kuwait on the Prairie: Can North Dakota Solve the Energy Problem?', *New Yorker*, 18 April 2011.

iii   R. Howarth, A. Ingraffea and T. Engelder, 'Should Fracking Stop?', *Nature*, 2011, no. 477, pp. 271–5.

iv   Lucy Lippard, *Undermining: A Wild Ride Through Land Use, Politics, and Art in the Changing West* (New Press, 2013), p. 103.

v   'Fracking Cover-Up Continues Groundwater Contamination Disaster in Pavillion, Wyoming', Western Organization of Resource Councils, 21 February 2019.

vi   Rob Jordan, 'Stanford Researchers Show Fracking's Impact to Drinking Water Sources', *Stanford News*, 29 March 2016.

vii   Robyn Vincent, 'Scientist Dominic DiGiulio's Work Illuminated How Fracking Affects People and Environment', Four Corners Public Radio, 24 December 2020.

viii   Naohiko Ohkouchi, Junichiro Kuroda and Asahiko Taira, 'The Origin of Cretaceous Black Shales: A Change in the Surface Ocean Ecosystem and Its Triggers', *Proceedings of the Japanese Academy, Series B Physical and Biological Sciences*, 21 July 2015, vol. 91, no. 7, pp. 273–91.

ix   Ibid.

x   Cody Cottier, 'The Devonian Extinction: A Slow Doom That Swept Our Planet', *Discover* , 23 January 2021.

xi   Lippard, *Undermining*, p. 164.

EMERALD

i   Quoted in Elisa Vázquez de Gey, *La Princesa de Kapurthala* (Planeta, 2008), pp. 139–40 (my translation).

ii   Joanna Whalley and Geoffrey Munn, 'Examining Jewellery in Detail: The Al Thani Collection', *V&A blog*, 2015.

iii   Elisa Vázquez de Gey, kapurthalaprincess.com

iv   Ibid.

v   Joanna Hardy, 'The Jeweller's Art', in Jonathan Self, Joanna Hardy, Franca Sozzani and Hettie Judah, *Emerald: Twenty-one Centuries of Jewelled Opulence and Power* (Violette Editions/ Thames & Hudson, 2013), p. 103.

vi   Vázquez de Gey, *La Princesa*.

vii   Jonathan Self, 'The Story of Emeralds', in Self, Hardy, Sozzani and Judah, *Emerald*, p. 259.

viii   Ibid., p. 264.

ix   *Maharajas & Mughal Magnificence*, sale catalogue (Christies, 2019), lot 132: 'A Belle Époque Emerald And Diamond Brooch'.

x   Vázquez de Gey, *La Princesa*.

MALACHITE

i   Preston Remington, 'The Story of a Malachite Vase', *Metropolitan Museum of Art Bulletin* , February, 1945, New Series, vol. 3, no. 6, pp. 142–5 (Metropolitan Museum of Art, 1945).

ii   James Jackson Jarves, 'The San Donato Art Sale', *New York Times*, 25 March 1880.

iii   Remington, 'Story of a Malachite Vase'.

iv   Ibid.

v   'The Demidoffs-Their Origin', *New York Times*, 19 January 1879.

vi   Daniel Russell, 'The Demidoff Malachite Mine, Nizhne-Tagil'skoye, Russia', 26 January 2008, Mindat.org

vii   Wolfram Koeppe, 'Monumental Vase Lapidary Work: Early 19th Century; Pedestal and Mounts: 1819', 2008, Metropolitan Museum of Art (online catalogue at metmuseum.org).

viii   Yelena V. Grant, '"Russian Mosaic" and its Italian Connection: Malachite in the Decorative Arts in the 1780s–1800s', thesis (Corcoran College of Art and Design, 2011), p. 23.

ix   Ibid., p. 67.

MARBLE

i   John McPhee, *Annals of the Former World* (Farrar, Straus and Giroux, 1998), pp. 519–21.

ii   I am greatly indebted to the remarkable scholarship of Fabio Barry, whose *Painting in Stone: Architecture and the Poetics of Marble from Antiquity to the Enlightenment* (Yale University Press, 2020) I have drawn on rather more often in this text than it would be elegant to admit to.

iii   Richard Fortey, *The Earth: An Intimate History* (HarperCollins, 2004), p. 299.

iv   Barry, *Painting in Stone.*

v   Ibid.

vi   gymnasium from *gumnoi*, nakedness.

vii   Richard Sennett, *Flesh and Stone* (Faber and Faber 1994/ Penguin 2002), p. 33.

viii   Sennett, *Flesh and Stone*, pp. 35–51.

ix   C. Suetonius Tranquillus, *Divus Augustus*, Ch. 29, trans. by Alexander Thomson, 1889.

x   It was described as *candidus* – bright, lucid, transparent – from which we derive the English term candid and (ironically) candidate.

xi   Barry, *Painting in Stone*.

xii   B. Burrell, 'Phrygian for Phrygians: Semiotics of "Exotic" Local Marble', *Interdisciplinary Studies on Ancient Stone: Proceedings of the IX ASMOSIA Conference* (Institut Català d'Arqueologia Clàssica, 2009), p. 782.

xiii   *The Odes Of Horace*, Book III, Ode I: 'On Contentment', trans. by Christopher Smart, 1767.

xiv   Barry, *Painting in Stone*, p. 79.

NEPHRITE

i   Chris Gosden, *The History of Magic: From Alchemy to Witchcraft, from the Ice Age to the Present* (Viking, 2020), pp. 112–13.

ii   Robert Bagley, 'Shang Archaeology', in Michael Loewe and Edward L. Shaughnessy (eds), *The Cambridge History of Ancient China from the Origins of Civilization to 221 B.C.* (Cambridge University Press, 1999), p. 195.

iii   Elizabeth Lyons, 'CHINESE JADES: The Role of Jade in Ancient China: An Introduction to a Special Exhibition at the University Museum', Penn Museum, PA, 1978 (published online, penn.museum).

iv   'Ai Weiwei on a Shang Dynasty Figure from the Tomb of Fu Hao', quoted in Jori Finkel, *It Speaks to Me: Art that Inspires Artists* (DelMonico Books/ Prestel, 2019), excerpted for Columbia Forum (published online, college.columbia.edu).

v   Burial Ensemble of Dou Wan, Metropolitan Museum of Art (online catalogue, metmuseum.org).

vi   Lyons, 'CHINESE JADES'.

vii   Jessica Rawson, 'The Eternal Palaces of the Western Han: A New View of the Universe', *Artibus Asiae*, 1999, vol. 59, nos 1/2, p. 14.

viii   Ibid., p. 50.

OLD RED SANDSTONE

i   J. Ussher, *The Annals of the World* (1658), 'The Epistle to the Reader'. Technically Ussher described the creation as happening in the night before the first day – which would actually be 22 October. It was not Ussher but the English parson and naturalist Sir John Lightfoot, who suggested 9 a.m. to be the time man was created.

ii   Henry Faul and Carol Faul, *It Began with A Stone: A History of Geology from the Stone Age to the Age of Place Tectonic* (Wiley Interscience, 1983), pp. 96–7.

iii   John Playfair, quoted in John McPhee, *Annals of the Former World* (Farrar, Straus and Giroux, 1998) pp. 78–9.

iv   The British have a small doughy cake called a scone, which has opened the stone up to much hilarity. In Terry Pratchett's Discworld universe, there's a sacred piece of dwarf bread called the Scone of Stone on which the Low Kings of the Dwarfs sit. In the real world, the cake is pronounced to rhyme with gone, and the stone, 'scone' to rhyme with tune.

v   Grant Thomson, '20 Facts Revealed About the Stone of Destiny', Historic Environment Scotland, published online, 29 November 2016.

vi   P. W. Joyce, 'The Lia Fáil and the Westminster Coronation Stone', *Irish Monthly*, vol. 12, no. 133 (Irish Jesuit Province, 1884), p. 325.

vii   N. J. Fortey, E. R. Phillips, A. A. McMillan and M. A. E. Browne, 'A Geological Perspective on the Stone of Destiny', 1 November 1998, *Scottish Journal of Geology*, vol. 34, pp. 145–52.

RUBY

i   George F. Bass, *World Beneath the Sea*, ed. Robert L. Breeden (National Geographic Society, 2nd edn, 1973), p. 130.

ii   Cormac F. Lowth, 'Finds of the Spanish Armada', *Dublin Historical Record*, spring 2004, vol. 57, no. 1 (Old Dublin Society), pp. 24–37.

iii   At this time the Spanish had control of the territory of Portugal.

iv   Lowth, 'Spanish Armada', pp. 24–37.

v   Ibid., p. 29.

vi   Doménikos Theotokópoulos known as El Greco, *Don Alonso Martínez de Leyva* (1580), Montreal Museum of Fine Arts.

vii   Laurence Flanagan, 'The Irish Legacy of the Spanish Armada', *Archaeology Ireland*, winter 1988, vol. 2, no. 4 (Worldwell Ltd), p. 147.

viii   Winifred Glover, 'The Gold of The *Girona*' (published online, lalampadina.net, 2013).

ix   Gaspero Balbi, *Voyage to Pegu, and Observations There* (c. 1583).

x   *The Etymologies of Isidore of Seville*, trans. by Stephen A. Barney, W. J. Lewis, J. A. Beach and Oliver Berghof (Cambridge University Press, 2006), p. 326.

SAPPHIRE

i   Seth Mydans, 'Marcos Flees and is Taken to Guam; U.S. Recognizes Aquino as President', *New York Times*, 26 February 1986.

ii   Bernard Gwertzman, 'Marcos Family Jewelry Brought to Hawaii is Put At $5 Million to $10 Million', *New York Times*, 8 March 1986.

iii   Lauren Greenfield, *The Kingmaker*, Showtime, 2019.

iv   Mark Fineman, 'Philippine Poor Gape at Marcos Palace Riches', *Los Angeles Times*, 15 March 1986.

v   Fox Butterfield, 'Manila is Freezing all Marcos Assets', *New York Times*, 13 March 1986.

vi   Timothy Walker, 'Christie's Auction House says Imelda Marcos Jewellery is "Fit for Royalty"', *Philippines Lifestyle News*, 26 November 2015.

vii   Fox Butterfield, 'In Manila Palace: Silk Dresses, 6,000 Shoes', *New York Times*, 9 March 1986.

viii   Judith Goldstein, 'Lifestyles of the Rich and Tyrannical', *American Scholar*, spring 1987, vol. 56, no. 2, p. 239.

ix   Joanna Hardy, *Sapphire* (Violette Editions/ Thames & Hudson, 2021), pp. 135–7.

x   Greenfield, *The Kingmaker*.

xi   These included Imelda's New York secretary, Vilma H. Bautista, whose name appeared on the purchase papers for four apartments in the Olympic Tower on Fifth Avenue, on the documents for an apartment on the East River, on a $1.43 million bill for emeralds, rubies and diamonds from Bulgari, and on the property tax bill for a substantial waterfront estate on Long Island.

xii   Material from Pio Abad and Frances Wadsworth Jones based on conversation and correspondence. Their work brought this story to my attention and I am deeply grateful to them both for their generosity in sharing their research material and time with me.

xiii   Sir Arthur Conan Doyle, 'The Adventure of the Blue Carbuncle' (1892), *The Complete Sherlock Holmes* (Doubleday, 1930), p. 249.

SACRED STONES

i   Edward Abbey, *Desert Solitaire* (McGraw-Hill, 1968) p. 231

AMETHYST

i   Barbara G. Walker, *Book of Sacred Stones: Fact and Fallacy in the Crystal World* (Harper & Row, 1989), pp. 92–4.

ii   Plutarch, *Quaestiones Convivales*, 3.1.

iii   Nonnus, *Dionysiaca*, Book XII, trans. W.H.D. Rouse.

iv   Quoted in Jean Braybrook, *Remy Belleau et l'art de guérir* (Versita, 2013) (my translation).

v   Daniel Trinchillo of Mardani Fine Minerals in New York, quoted in Rina Raphael, 'Is There a Crystal Bubble? Inside The Billion-Dollar "Healing" Gemstone Industry', *Fast Company*, 5 May 2017.

vi   Tess McClure 'Dark Crystals: The Brutal Reality Behind a Booming Wellness Craze', *Guardian*, 17 September 2019.

vii   Fábio Lima, 'Uma vinícola dentro de uma mina de pedras preciosas no Rio Grande Do Sul', *InTrip*, 2 February 2015.

CAIRNGORM

i   Benjamin Woollet, *The Queen's Conjuror: The Life and Magic of Dr Dee* (HarperCollins, 2001), p. 3.

ii   Chris Gosden, *The History of Magic: From Alchemy to Witchcraft: from the Ice Age to the Present* (Viking, 2020), p. 381.

iii   Woollet, *Queen's Conjuror*, p. 91.

iv   Ibid., p. 168.

v   Ibid., p. 56.

vi   John Dee's diary, quoted in Charles Nicholl, 'The Last Years of Edward Kelley, Alchemist to the Emperor', *London Review of Books*, 19 April 2001.

vii   Dee is considered a model for Shakespeare's Prospero, who summons the spirit Ariel.

viii   Deborah E. Harkness, 'Shows in the Showstone: A Theater of Alchemy and Apocalypse in the Angel Conversations of John Dee (1527–1608/9)', *Renaissance Quarterly*, winter 1996, vol. 49, no. 4, p. 716.

ix   John Dee's diary, 21 November 1582, quoted in Woollet, *Queen's Conjuror*, p. 189.

x   Quoted in Nicholl, 'Last Years of Edward Kelley'.

xi   Ibid.

xii   John Maynard Keynes, 'Newton the Man'; a lecture written by Keynes and delivered posthumously on his behalf to the Royal Society, July 1946.

CINNABAR

i   Anita Chowdry, 'More Than the Colour Red: The Unspoken Symbolism of Cinnabar Pigment in Indian Painting', paper for the international conference, 'Visions of Enchantment: Occultism, Spirituality and Visual Culture', University of Cambridge, 2014, p. 14.

ii   The elements sulphur and mercury were not precisely alchemical sulphur and alchemical mercury, but stood in as representative symbols for them. It's all very complicated, deliberately so.

iii   Chowdry, 'The Colour Red', p. 2.

iv   Ibid., p. 13.

v   Ibid., p. 10.

vi   David Gordon White, *The Alchemical Body: Siddha Traditions in Medieval India* (University of Chicago Press, 2012), p. 187.

vii   The tradition dates to the founding of the Shrinathji Temple in Nathdwara, Rajasthan.

viii   I am indebted to Desmond Peter Lazaro, *Materials, Methods & Symbolism in the Pichhvai Painting Tradition of Rajasthan* (Mapin Publishing, 2005), a book I draw on throughout this text. And to the painter Olivia Fraser who directed me to it.

ix   Ibid., p. 55.

GLOBIGERINA LIMESTONE

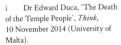

i   Dr Edward Duca, 'The Death of the Temple People', *Think*, 10 November 2014 (University of Malta).

ii   Katya Stroud, 'Of Giants and Deckchairs: Understanding the Maltese Megalithic Temples', in Caroline Malone and David Barrowclough (eds), *Cult in Context: Reconsidering Ritual in Archaeology* (Oxbow Books, 2010), pp. 16–22.

iii   Marija Gimbutas, *The Living Goddesses* (University of California Press, 1999), 'Stone Temples of Malta' pp. 93–7.

iv   Caroline Malone, 'Ritual Space and Structure – the Context of Cult in Malta and Gozo', in Malone and Barrowclough (eds) *Cult in Context*, pp. 23–34.

v   Anthony Bonanno, Caroline Malone, Tancred Gouder et al., 'The Death Cults of Prehistoric Malta', *Scientific American, Special Edition*, vol. 15, no. 1s, pp. 14–23, January 2005.

GRANITE

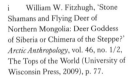

i   William W. Fitzhugh, 'Stone Shamans and Flying Deer of Northern Mongolia: Deer Goddess of Siberia or Chimera of the Steppe?' *Arctic Anthropology*, vol. 46, no. 1/2, The Tops of the World (University of Wisconsin Press, 2009), p. 77.

ii   William W. Fitzhugh, 'The Mongolian Deer Stone-Khirigsuur Complex: Dating and Organization of a Late Menagerie', *Current Archaeological Research in Mongolia* (Vor- und Fruhgeschichtliche Archeologie, Rheinische Friedrich-Wilhelms-Universitat, 2009), p. 185.

iii   William Timothy Treal Taylor, Julia Clark, Jamsranjav Bayarsaikhan et al., 'Early Pastoral Economies and Herding Transitions in Eastern Eurasia', *Nature* Scientific Reports, 2020, p. 2.

iv   Ibid., p. 3.

v   Antoine Fages, Kristian Hanghøj, Naveed Khan et al., 'Tracking Five Millennia of Horse Management with Extensive Ancient Genome Time Series', *Cell*, 30 May 2019, vol. 177, no. 6, pp. 1,419–35.

vi   Chris Gosden, *The History of Magic: From Alchemy to Witchcraft, from the Ice Age to the Present* (Viking, 2020), p. 150.

vii   Fitzhugh, 'Stone Shamans', p. 77.

viii   Taylor, Clark, Bayarsaikhan et al., 'Early Pastoral Economies', p. 3.

ix   Claudia Chang and Rebecca Roberts (eds), *Gold of the Great Steppe: People, Power and Production* (Paul Holberton, 2021).

x   Esther Jacobson, *The Deer Goddess of Ancient Siberia: A Study in the Ecology of Belief* (E. J. Brill, 1993), pp. 3–4.

JADEITE

i   The *Popol Vuh* was set down by the elite of the Quiché Maya people in Guatemala prior to the arrival of the Spanish in 1524, and was kept preserved and hidden for centuries. The sacred stories long pre-date the text itself.

ii   Stephen Houston, 'A Beauty that Cannot Die', 2014, blog.yalebooks.com

iii   Lisa J. Lucero, 'The Politics of Ritual: The Emergence of Classic Maya', *Current Anthropology*, August/October 2003, vol. 44, no. 4 (University of Chicago Press on behalf of the Wenner–Gren Foundation for Anthropological Research, 2003), pp. 531–2.

iv   Michael Coe (1988), quoted in Karl A. Taube, 'The Symbolism of Jade in Classic Maya Religion', *Ancient Mesoamerica*, 2005, vol. 16, no. 1 (Cambridge University Press), pp. 23–50 at pp. 30–31.

v   Houston, 'Beauty that Cannot Die'.

vi   A city-state in what is now southern Mexico.

vii   U.Pasqualini and M.E.Pasqualini, *Treatise of Implant Dentistry: The Italian Tribute to Modern Implantology* (Ariesdue, 2009) Ch. 1: The History Of Implantology.

viii   *Popol Vuh: Sacred Book of the Quiché Maya People*, trans. and commentary by Allen J. Christenson (University of Oklahoma Press, 2003) ll. 985–1,188: The Defeat of Seven Macaw.

JET

i   Livy, *Ab urbe condita*, 29.14, (c.27–9 BCE). Later accounts describe the boat running onto a sandbank and Claudia Quinta miraculously towing the craft to safety with her girdle.

ii   Marika Rauhala, 'Devotion and Deviance: The Cult of Cybele and the Others Within', in Maijastina Kahlos (ed.), *The Faces of the Other: Religious Rivalry and Ethnic Encounters in the Later Roman World* (Cursor Mundi, 2012), p. 52.

iii   Ibid., pp. 56–7.

iv   The ancient kingdom of Phrygia was located in present-day Anatolia, Turkey.

v   H. Cool, 'The Catterick Gallus', *Lucerna*, July 2002, no. 24, pp. 18–21.

vi   Sarah Steele, 'What is Whitby Jet?' Ebor Jetworks Blog, published online 14 August 2020. Steele is a lapidary and PhD candidate who has exploded the popular myth that jet is the petrified remains of monkey puzzle-type trees: she has identified the remains of at least six different tree species in jet, none of them *Araucaria araucana*, the monkey puzzle.

vii   Pliny, *Naturalis Historia* (The Natural History) (c. 77 CE), Book 36, section 141.

viii   Christopher A. Faraone, *The Transformation of Greek Amulets in Roman Imperial Times* (University of Pennsylvania Press, 2018).

ix   Lindsay Allason-Jones, *Roman Jet in the Yorkshire Museum* (Yorkshire Museum, 1996), p. 17.

x   Rauhala, 'Devotion and Deviance', pp. 75–6.

PELE'S HAIR

i   E.S.Craighill Handy and Mary Kawena Pukui, *The Polynesian Family System in Ka-'U, Hawai'i* (Wellington, N.Z., Polynesian Society, 1958), p. 29; ku'ualoha ho'omanawanui, *Voices of Fire: Reweaving the Literary Lei of Pele and Hi'iaka* (University of Minnesota Press, 2014), p. xxiv.

ii   Richard Fortey, *The Earth: An Intimate History* (HarperCollins, 2004), p. 75.

iii   ho'omanawanui, *Voices of Fire*, p. xxxvii.

iv   David A.Chang, *The World and All the Things upon It: Native Hawaiian Geographies of Exploration* (University of Minnesota Press, 2016), pp. 218–24.

v   Handy and Pukui, *Polynesian Family System*, p. 28.

vi   This retelling is based on a version given by Hawaiian scholar ku'ualoha ho'omanawanui in her *Voices of Fire*, pp. 26–9.

vii   Julie Cart, 'Hawaii's Hot Rocks Blamed by Tourists for Bad Luck', *Los Angeles Times*, 17 May 2001.

SARSEN

i   Alex Bayliss, Alasdair Whittle and Michael Wysocki, 'Talking About my Generation: The Date of the West Kennet Long Barrow', *Cambridge Archaeological Journal*, vol. 17, no. S1, p. 86.

ii   Ibid.

iii   Mark Gillings, Joshua Pollard and Kristian Strutt, 'The Origins of Avebury', *Antiquity*, April 2019, vol. 93, no. 368, pp. 359–77, published online by Cambridge University Press 10 April 2019.

iv   UNESCO, 'Stonehenge, Avebury and Associated Sites' (published online, whc.unesco.org).

v   'Old English (Anglo-Saxon) Words and Influences', 2014, Tolkien Society (published online, tolkiensociety.org).

vi   *The Diary of Samuel Pepys: Daily Entries from the 17th Century London Diary*, Monday 15 June 1668 (accessed online, pepysdiary.com).

vii   David J. Nash, T. Jake R. Ciborowski, J. Stewart Ullyott et al., 'Origins of the Sarsen Megaliths at Stonehenge', *Science Advances*, 29 July 2020, vol. 6, no. 31.

viii   Mark Gillings and Joshua Pollard, 'Making Megaliths: Shifting and Unstable Stones in the Neolithic of the Avebury Landscape', *Cambridge Archaeological Journal*, 2016, vol. 26, no. 4, p. 3.

ix   Jonathan R. Trigg, 'Re-examining the Contribution of Dr Robert Toope to Knowledge in Later Seventeenth-century Britain: Was He More than Just "Dr Took"?' in Julia Roberts, Kathleen Sheppard, Ulf R. Hansson and Jonathan R. Trigg (eds), *Communities and Knowledge Production in Archaeology*, (Manchester University Press, 2020), p. 207.

x   Jonathan Charles Goddard, 'Goddard's Dropps: A Paradox of the 17th Century Urology', *Urology News*, September/ October 2015, vol. 19, no. 6, online at Museum of Urology, BAUS (British Association of Urological Surgeons).

xi   Robert Wallis and Jenny Blain, 'The Sanctity of Burial: Pagan Views, Ancient and Modern', paper for the conference, 'Respect for Ancient British Human Remains: Philosophy and Practice', Manchester Museum, University of Manchester, 17 November 2006, p. 8.

xii   Blog post, 'An impression of West Kennet Long Barrow', Marc Rhodes-Taylor of the Druid Network.

TUFF

i   Jared Diamond, *Collapse: How Societies Choose to Fail or Survive* (Penguin, 2013), Ch. 2.

ii   Teuira Henry et al., *Voyaging Chiefs of Havai'i: Sixteen Narratives of Voyaging Migrations and Travels, from Around Polynesia* (Kalamaku/ Noio & Far Roads, 1995), 'Hotu Matua', published online 2020.

iii   Terri Cook and Lon Abbott, 'Travels in Geology: Easter Island's Enduring Enigmas', *Earth*, 23 March 2017.

iv   Jo Anne Van Tilburg with T. Ralston, 'Megaliths and Mariners: Experimental Archaeology on Easter Island' , in K.L. Johnson (ed.), *Onward and Upward: Papers in Honor of Clement W. Meighan* (Stansbury Publishing, 2005), pp. 279–306.

v   Forrest Wade Young, 'I Hē Koe? Placing Rapa Nui', *Contemporary Pacific*, 2012, vol. 24, no. 1 (University of Hawai`i Press), pp. 1–30.

vi   The remains of a leper colony endure on Rapa Nui, where it is regarded by many as a tool of colonial oppression, see Young 'I Hē Koe?' pp. 12–14.

vii   Diamond, *Collapse*.

viii   'Unearthing the Mystery of the Meaning of Easter Island's Moai: Rapanui People Likely Believed the Ancient Monoliths Helped Food Grow on the Polynesian Island, Study Reveals', University of California, ScienceDaily, 13 December 2019.

ix   Sarah Sherwood, Casey Barrier, José Miguel Ramírez-Aliaga et al., 'New Excavations in Easter Island's Statue Quarry: Soil Fertility, Site Formation and Chronology', *Journal of Archaeological Science*, 2019.

TURQUOISE

i   Marjorie Caygill, 'The Provenance of Three British Museum Turquoise Mosaics', 5 April 2010, Mexicolore.co.uk

ii   Sahagún worked with Nahua men on the history, *La historia general de las Cosas de Nueva España* – also known as the 'Florentine Codex' – for forty-five years, 1545–90. Sahagún had arrived in 'New Spain' in 1529, ten years after the encounter between Cortés and Moctezuma.

iii   The *Codex* was named after the viceroy of 'New Spain'. Its Nahuatl pictograms and Spanish text were put together in about 1541. They detail the history of the Aztec rulers and provide information about life in pre-conquest societies.

iv   Eric A. Powell, 'The Turquoise Trail', *Archaeology*, January/February 2005, vol. 58, no. 1 (Archaeological Institute of America), pp. 24–9.

v   I am indebted here to the oral history of the Diné told by Navajo historian Wally Brown.

vi   Peter N. Peregrine, 'Matrilocality, Corporate Strategy, and the Organization of Production in the Chacoan World', *American Antiquity*, January 2001, vol. 66, no. 1, (Cambridge University Press), p. 39.

vii   Frances Joan Mathien, 'The Organization of Turquoise Production and Consumption by the Prehistoric Chacoans', *American Antiquity*, January 2001, vol. 66, no. 1 (Cambridge University Press), p. 116.

viii   Powell, 'Turquoise Trail', p. 28.

ix   Colin Renfrew, 'Production and Consumption in a Sacred Economy: The Material Correlates of High Devotional Expression at Chaco Canyon', *American Antiquity*, January 2001, vol. 66, no. 1 (Cambridge University Press), p. 18.

x   Patricia L. Crown and W. Jeffrey Hurst, 'Evidence of Cacao Use in the Prehispanic American Southwest', *PNAS*, 17 February 2009, vol. 106, no. 7, (Proceedings of the National Academy of Sciences of the United States of America), pp. 2, 110–13.

STONES AND STORIES

i   J. R. R. Tolkien, 'On Fairy Stories', first delivered as the Andrew Lang lecture at the University of St Andrews, Scotland, on 8 March 1939, published in C. S. Lewis (ed.), *Essays Presented to Charles Williams* (Oxford University Press, 1947), accessed online.

ii   Paul Basu, 'Sacred Stone Axes on Benin Altars', *[Re:] Entanglements: Nigeria / Sierra Leone / Re-engaging with Colonial Archives in Decolonial Times*, published online 29 April 2020.

CALAVERITE

i   Since the vine can be found throughout the arid regions of Australia, there are other indigenous names: according to the indigenous edibles blog warndu.com, around Alice Springs it is called langkwe and in the Flinders Ranges it is myakka.

ii   superpit.com.au

iii   'On the Track', from the history of the Western Australia Goldfield, museum.wa.gov.au

iv   Sergey V. Streltsov, Valerii V. Roizen, Alexey V. Ushakov et al., 'Old Puzzle of Incommensurate Crystal Structure of Calaverite AuTe$_2$ and Predicted Stability of Novel AuTe Compound', *Proceedings of the National Academy of Sciences of the United States of America*, 2 October 2018, vol. 115, no. 40, pp. 9,945–50.

v   Richard Fortey, *The Earth* (HarperCollins, 2004), pp. 269–70.

CHRYSOBERYL

i   J. K. Huysmans, *Against Nature*, trans. Robert Baldick (Penguin, 1959), p. 54.

ii   Ibid., p. 55.

iii   Oscar Wilde, *The Picture of Dorian Gray* (1891) (World's Classics, Oxford University Press, 1974), p. 135.

iv   Ezekiel xxviii, 22, King James version.

v   Paul E. Desautels, *The Gem Kingdom* (A Ridge Press Book, Random House, 1976), p. 93.

vi   Victoria Mills, 'Dandyism, Visuality and the "Camp Gem": Collections of Jewels in Huysmans and Wilde', in L. Calè and P. Di Bello (eds), *Illustrations, Optics and Objects in Nineteenth-Century Literary and Visual Cultures* Palgrave Studies in Nineteenth-Century Writing and Culture (Palgrave Macmillan, 2010).

vii   Walter Horatio Pater, *The Renaissance* (1873), Conclusion.

viii   Quoted in Helen Andrews, 'Their Decadence and Ours', *First Things*, 14 October 2014.

DIAMOND

i   T. S. Eliot, quoted in Deirdre David (ed.), *The Cambridge Companion to the Victorian Novel* (Cambridge University Press, 2001), p. 179.

ii   Wilkie Collins, Preface to *The Moonstone* (1868).

iii   metmuseum.org/blogs/now-at-the-met/2018/indian-diamonds-benjamin-zucker-family-collection

iv   Ibid.

v   Alice Procter, 'Pitt's Diamond', *The Whole Picture* (Cassell, 2020), pp. 46–52.

vi   Ibid.

vii   Shuhita Bhattacharjee, 'The Insurgent Invasion of Anti-Colonial Idols in Late-Victorian Literature: Richard Marsh and F. Anstey', *English Literature in Transition, 1880–1920*, 2018, vol. 61, pp. 66–90.

viii   E. W. Streeter, *The Great Diamonds of the World: Their History and Romance* (G. Bell & Sons, 1882).

ix   Victoria Findlay, *Jewels: A Secret History* (Hodder & Stoughton, 2005), p. 349.

DOLERITE

i   In their song *Stonehenge*, which features in the film, *This is Spinal Tap* (1984).

ii   *Sacrilege* (2012).

iii   John Michell, *Megalithomania* (Thames & Hudson, 1982), p. 12.

iv   Ibid., p. 23.

v   A demon explained in the text as 'in nature partly of men and partly of angels and whenever they please assume human shapes, and lie with women,' Geoffrey of Monmouth, *History of the Kings of Britain* (c. 1136), trans. Aaron Thompson with revisions by J. A. Giles (In Parentheses Publications, 1999), p. 110.

vi   Ibid., pp. 133–6.

vii   Mike Parker Pearson, Josh Pollard, Colin Richards et al. 'The Original Stonehenge? A Dismantled Stone Circle in the Preseli Hills of West Wales', *Antiquity*, vol. 95, no. 379 (Cambridge University Press, 2021), pp. 85–103.

viii   Aubrey Burl, *The Stone Circles of Britain, Ireland and Brittany* (Yale University Press, 2000), p. 350.

ix   Ibid. p. 365.

x   Pearson, Pollard, Richards et al., 'Original Stonehenge?'.

xi   Paul Devereux and Jon Wozencroft, 'Stone Age Eyes and Ears: A Visual and Acoustic Pilot Study of Carn Menyn and Environs, Preseli, Wales, Time and Mind', *Journal of Archaeology, Consciousness and Culture*, 2014, vol. 7, no. 1, pp. 47–70.

xii   Ibid.

xiii   Thomas Hardy, *Tess of the D'Urbevilles* (1891), Phase the Seventh: Fulfilment/Ch. 58.

LAPIS LAZULI

i   *The Epic of Gilgamesh* here and throughout, loosely adapted from the Penguin Classics 1989 translation by Maureen Gallery Kovacs.

ii   A number of other versions have since been discovered.

iii   A Chaldean Christian, born in 1826 in Mosul, then part of the Ottoman empire.

iv   The translator was George Smith. The sensation was caused largely because of the story's account of a deluge, predating the first written version of the Old Testament.

v   Daniel Silas Adamson, 'The men who uncovered Assyria', *BBC News magazine*, 22 March 2015.

vi   Both diadem and sculptures are in the collection of the Penn Museum, Philadelphia.

vii   Georgina Herrmann, 'Lapis Lazuli: The Early Phases of Its Trade', *Iraq*, vol. 30, no. 1 (British Institute for the Study of Iraq, 1968), p. 21.

viii   Ibid.

ix   This is not the end of the *Epic* – Gilgamesh, tormented by the prospect of his own death, grows wild like Enkidu and crosses mountains and seas to seek the last survivor of the great deluge and learn the secret of immortality. He learns instead that he is mortal but that humanity will endure, and returns to Uruk a better man and a better king. Lapis lazuli persists as an emblem along his route: as a ritual bowl, the leaves of enchanted trees and as beads of memory.

MOLDAVITE

i   *The Adventures of Superman* series was written by George Putnam Ludlam.

ii   Grant Morrison, *Supergods* (Vintage, 2011), p. 72.

iii   The finding of Superbaby owes no small debt to the story of Moses.

iv   *Superman* #61.

v   Morrison, *Supergods*, p. 73.

vi   When first inked, kryptonite was in fact red.

vii   Paul Fairchild, 'Kryptonite Is Crap: The weird, dumb history of Superman's ill-conceived vulnerability', *Atlantic*, 20 June 2013.

viii   Rebekah Harding, 'WitchTok Is Obsessed With This Life-Changing Piece of Glass', *Cosmopolitan*, published online 30 April 2021.

ix   Ibid.

x   Mariella Bucci quoted in Becki Parker, 'Moldavite is the Healing Crystal TikTokers are Saying Ruins Lives', *Cosmopolitan*, published online 11 August 2021.

xi   *Sleeping With Moldavite: A Beginner's Guide*, Yes Dirt (published online, yesdirt.com).

xii   Chermaine Chee, 'Moldavite: Meaning, Healing Properties & Uses Of This Powerful Crystal', *Truly* 20 September 2021.

xiii   Parker, 'Moldavite is the Healing Crystal'.

xiv   'South Bohemia Faces Growing Wave of Illegal Moldavite Miners', Radio Prague International (published online, english.radio.cz) 25 January 2021.

MOON ROCK

i   Robert M. Hazen, *The Story of Earth* (Penguin, 2012), pp. 32–3.

ii   Ibid., p. 35.

iii   Ibid., pp. 38–9.

iv   William K. Hartmann And Donald R. Davis, 'Satellite-Sized Planetesimals and Lunar Origin', *ICARUS* 24 (Academic Press, 1975), pp. 504–515.

v   Conference on the Origin of the Moon, Kailua-Kona, Hawai'i, 13–16 October 1984, Reminiscences by G. Jeffrey Taylor, December 31, 1998, psrd.hawaii.edu/Dec98/reminisces.html

OPAL

i   John Francis Dodson, 'Minnie Berrington – Summary of her Life as Told in Stones of Fire', Johno's Opals blog, 23 August 2013 (johnosopals.com/blog).

ii   Ibid.

iii   Russell Shor, 'The Real Gemology of Ethiopian Opals in "Uncut Gems"', Gemological Institute of America (published online, gia.edu), 22 January 2020.

iv   Stuart Wattison, 'Minnie Berrington – Andamooka's First Postmistress', Johno's Opals blog, 4 October 2018 (johnosopals.com/blog).

v   Quoted in John Piggott, 'Pioneering Minnie – Lover of Opals and Outback – Celebrated in Upcoming Film', *Senior*, 18 December 2021.

vi   Wattison, 'Minnie Berrington'.

PHONOLITE PORPHYRY

i   Richard Erdoes and Alfonso Ortiz (eds), *American Indian Myths and Legends* (Pantheon Books, 1984), pp. 227–9.

ii   Ibid., p. 227.

iii   'Devils Tower: First Explorers', information originally drafted by Ray H. Mattison, historian for the National Park Service, 1955, nps.gov

iv   Mike Smith, 'Interview with Joe Alves', 2010, mediamikes.com

v   This formation is called a Maar-diatreme volcano and was proposed by the geologist Prokop Závada and colleagues in 2015, in P. Závada, P. Dědeček, J. Lexa and G. R. Keller, 'Devils Tower (Wyoming, USA): A Lava Coulée Emplaced into a Maar-diatreme Volcano?' *Geosphere*, April 2015, vol. 11, pp. 354–75.

PUMICE

i   Paul E. Desautels, *The Mineral Kingdom* (Paul Hamlyn, 1969), p. 129.

ii   Ice Cube, 'What is a Pyroclastic Flow?' on the album *Raw Footage* (Lench Mob Records/ EMI, 2008).

iii   'The Hazards of Pyroclastic Flows', 31 October 2014, National Geographic online.

iv   The scale in fact goes up to 'ultra Plinian'.

v   Pliny the Younger, *Letters*, LXV. To Tacitus (The Harvard Classics, 1909–14), translated by William Melmoth.

vi   Alwyn Scarth, *Volcanoes: An Introduction* (CRC Press, 2004), p. 79.

vii   The cause was previously thought to have been asphyxiation. See Giuseppe Mastrolorenzo, Pierpaolo Petrone, Lucia Pappalardo et al., 'Lethal Thermal Impact at Periphery of Pyroclastic Surges: Evidences at Pompeii', 2010, *PLos ONE*, vol. 5, no. 6.

viii   Sissel Tolaas, interview with the author, 19 November 2021.

SPINEL

i   Joanna Hardy, *Ruby* (Violette Editions/ Thames & Hudson, 2017), p. 34.

ii   Ibid., p. 38.

iii   Max Stern, *Gems: Facts, Fantasies, Superstitions, Legends* (Max Stern & Co., 1946), quoted in Barbara G. Walker, *Book of Sacred Stones: Fact and Fallacy in the Crystal World* (Harper & Row, 1989), p. 179.

iv   Paul Desautels, *The Gem Kingdom* (Random House, 1977), pp. 65, 119.

v   I have included this bit of myth busting because the episode characterises much of the research process for this book: the history of jewellery seems composed largely of concoctions and fabrications. I apologise for any disappointment caused.

vi   Quoted in Jack Ogden, 'The Black Prince's Ruby: Investigating the Legend', *Journal of Gemmology*, 2020, vol. 37, no. 4, p. 361.

vii   Ibid., pp. 360–71. I am grateful to Joanna Hardy for bringing this essay to my attention. Had she not, mine would have been yet one more name on the long list of writers who have, without questioning it, retold the legend of the Black Prince Ruby.

SHAPES IN STONE

i   Dore Ashton, *Noguchi East and West* (University of California Press, 1992), p. 100.

ii   Matt Kirsch, 'The Search for Stones', in Dakin Hart (ed.), *Museum of Stones: Ancient and Contemporary Art at the Noguchi Museum* (Isamu Noguchi Foundation, New York, 2016), p. 20.

iii   Quoted in Ashton, *Noguchi East and West*, p. 9.

AQUAMARINE

i   Abigail Eisenstadt 'How the World's Largest Aquamarine Gem Came to Be', *Smithsonian Voices*, 16 March 2021.

ii   Brian Vastag, 'The Dom Pedro Aquamarine's Long and Winding Path to the Smithsonian', *Washington Post*, 2 December 2012.

iii   Author interview with Kieran McCarthy, 3 December 2021.

iv   Vastag, 'The Dom Pedro'.

BASALT

i   Robert Smithson, 'The Spiral Jetty', in Jack Flam (ed.), Robert Smithson the Collected Writings (University of California Press, 1996), p. 145.

ii   Ibid., p. 146.

iii   James Housefield, 'Sites of Time: Organic and Geologic Time in the Art of Robert Smithson and Roxy Paine', Cultural Geographies, vol. 14, no. 4 (Sage Publications, 2007), pp. 537–6.

iv   Richard Fortey, The Earth: An Intimate History (HarperCollins, 2004), p. 80.

v   Danielle Hall, 'Making a Mark on the Ocean Floor', Smithsonian Ocean, July 2016.

vi   Fortey, The Earth, p. 89.

vii   Robert M. Hazen, The Story of Earth (Penguin, 2012), pp. 116–17.

viii   Ibid.

ix   Ronald Mason and Arthur Raff, 'Magnetic Survey off the West Coast of North America, 40° N. Latitude to 52° N. Latitude', GSA Bulletin vol. 72, no. 8 (The Geological Society of America, 1961), pp. 1,267–70.

x   Robert Smithson, 'The Artist as Site-Seer; or a Dintorphic Essay', in Flam (ed.), Robert Smithson the Collected Writings, p. 342.

xi   Robert Smithson, 'A Sedimentation of the Mind: Earth Projects', Artforum (1968), in Flam (ed.), Robert Smithson the Collected Writings, p. 105.

xii   In his essay on Spiral Jetty, Smithson wrote: 'How to get across the geography of Gondwanaland, the Austral Sea, and Atlantis became a problem. Consciousness of the distant past absorbed the time that went into the making of the movie. I needed a map that would show the prehistoric world as coextensive with the world I existed in.' The Spiral Jetty film would include footage of dinosaurs, filmed at the American Museum of Natural History in New York.

xiii   Housefield, 'Sites of Time', p. 541.

xiv   Smithson, 'A Sedimentation of the Mind', p. 110.

CHALCEDONY

i   Ivan Stebakov, Paperweight, ca 1866–7, V&A Museum, London.

ii   'Colors of the Universe: Chinese Hardstone Carvings', Metropolitan Museum of Art (2016).

iii   Claire Voon. 'Mouthwatering Qing Dynasty Sculpture of Braised Pork Belly Leaves Asia for First Time', Hyperallergic, 23 August 2016.

iv   Jori Finkel, 'From Taipei, a Feast for the Eyes for a Time Before Instagram', New York Times, 27 May 2016.

v   Sarah Fritsche, 'World-Famous Pork Belly Stone Arrives in SF', San Francisco Chronicle, 14 June 2016.

vi   Isaac Yue, 'The Comprehensive Manchu–Han Banquet: History, Myth, and Development', Ming Qing Yanjiu, University of Hong Kong, published online 14 November 2018.

vii   Thomas S. Elias, 'Chinese Food Stones (shiwu): Overview of Small Stones Resembling Different Foods', published online by Viewing Stone Association of North America, April 2014.

CHALK

i   Maurie Leblanc, L'Aiguille creuse (1909).

ii   In fact, he was born in Liège in 1783, which then was in the Austrian Netherlands.

iii   C. S. Harris, 'Chalk Facts', geologyshop.co.uk

iv   Richard Fortey, The Hidden Landscape (Jonathan Cape, 1993), p. 217.

v   Mark A. Richards, Walter Alvarez, Stephen Self et al., 'Triggering of the largest Deccan Eruptions by the Chicxulub Impact', GSA Bulletin, 2015, vol. 127, nos 11–12), pp. 1,507–20.

vi   Emily Cleaver, 'Against All Odds, England's Massive Chalk Horse Has Survived 3,000 Years', Smithsonian magazine, 6 July 2017.

vii   Thomas Baskerville, 'The Description of Towns, on the Road from Faringdon to Bristow and Other Places'(1681), quoted in W. J. A. C. J. C.-B. Portland, J. Nalson, R. Harley et al. (eds), the Manuscripts of His Grace the Duke of Portland: Preserved at Welbeck Abbey (Eyre & Spottiswoode for H. M. Stationery Office, 1891–1931), vol. 2, pp. 297–8.

viii   For a deep dive into the local folklore and history of the Giant, read Rebecca Mead's fantastic 'Letter from England' on 'The Mysterious Origins of the Cerne Abbas Giant', New Yorker, 12 May 2021.

ix   John Julius Norwich on Iris Tree, interview for Web of Stories: Life Stories of Remarkable People, 19 June 2018.

x   Virginia, Marchioness of Bath, Daily Telegraph, 24 September 2003.

xi   Dorset Evening Echo, 27 April 1981, p. 9, quoted in Rodney Castleden, The Cerne Giant (Dorset Publishing, 1996) p. 36, with thanks to Rebecca Mead.

xii   'Looking After the Cerne Giant', nationaltrust.org.uk

xiii   'National Trust Archaeologists Surprised by Likely Age of Cerne Abbas Giant', nationaltrust.org.uk

GYPSUM

i   Edwin Heathcote, 'The New National Museum of Qatar is a Desert Rose of Mutant Scale', Financial Times, 28 March 2019.

ii   David Bressan, 'The Largest Crystals Ever Discovered are at Risk of Decay', Forbes, 5 July 2018.

iii   The Mohs scale of mineral hardness ranges from 1 – talc, to 10 – diamond. The mineral at each level of the scale is hard enough to mark the mineral on the level below. A human fingernail is 2.5 on the Mohs scale, hard enough to scratch a mark into gypsum.

iv   See also: Lapis Lazuli , p. 136

v   Thutmose [?], 'Bust of Nefertiti' (1345 BCE), Neues Museum, Berlin.

vi   Haniya Rae, 'An Exciting History of Drywall', *Atlantic*, 29 July 2016.

vii   B. Szostakowski, P. Smitham and W. S. Khan, 'Plaster of Paris – Short History of Casting and Injured Limb Immobilization', *Open Orthopaedics Journal*, 2017, vol. 11, pp. 291–6.

viii   Denis Diderot, 'Observations sur la sculpture et sur Bouchardon', quoted in Eckart Marchand, 'Apostles of Good Taste? The Use and Perception of Plaster Casts in the Enlightenment', *Journal of Art Historiography*, December 2021, no. 25 (University of Birmingham), p. 4.

ix   Ibid., p. 2.

x   'A Short History of Plaster Casts', Cornell Collections of Antiquities, Cornell University Library, (published online, antiquities.library.cornell.edu).

xi   Mary Beard, 'Casts and Cast-offs: The Origins of The Museum of Classical Archaeology', *Proceedings of the Cambridge Philological Society*, 1993, no. 39 (Cambridge University Press), p. 8.

xii   Bruno Latour and Adam Lowe 'The Migration of the Aura or How to Explore the Original Through its Fac Similes', in Thomas Bartscherer and Roderick Coover (eds), *Switching Codes* (University of Chicago Press, 2010).

xiii   Valentina Risdonne, Charlotte Hubbard, Victor Hugo López Borges and Charis Theodorakopoulos, 'Materials and Techniques for the Coating of Nineteenth-century Plaster Casts: A Review of Historical Sources', *Studies in Conservation*, published online 17 January 2021.

xiv   Beard, 'Casts and Cast-offs', pp. 20–1.

xv   Latour and Lowe, 'Migration of the Aura'.

LINGBI

i   The story is in Pu Songling's multi-volume work, *Liaozhai's Stories of the Strange (Liaozhai Zhiyi)*, written between 1618 and 1714.

ii   Maggie Keswick, *The Chinese Garden: Art History & Architecture* (Academy Editions, / St Martin's Press, 1978/86), p. 155.

iii   Kemin Hu, *The Spirit of Gongshi* (Art Media Resources, 1999), p. 21.

iv   Robert D. Mowry (ed.), *Worlds Within Worlds: The Richard Rosenblum Collection of Chinese Scholars' Rocks* (Harvard University Art Museums, 1997), p. 21.

v   Kemin Hu, *Spirit of Gongshi*, p. 24.

vi   James Blades, *Percussion Instruments and Their History* (Bold Strummer, 1992), p. 119.

vii   Hugh T. Scogin, Jr., 'A Note on Lingbi', in Mowry (ed.), *Worlds Within Worlds*, p. 48.

ONYX

i   Roger Caillois, *The Writing of Stones* (published in 1970 as *L'Ecriture des pierres*), trans. by Barbara Bray (University of Virginia, 1985), p. 64.

ii   Marina Warner, 'The Writing Of Stones: Roger Caillois's imaginary logic', *Cabinet*, spring 2008.

iii   Cryptocrystalline stones are made up of a mass of tiny crystals: hence an individual quartz crystal is icy and transparent, but cryptocrystalline quartz is waxy and translucent.

iv   Caillois, *Writing of Stones*, pp. 64–7.

v   Ibid., p. 70.

vi   Ibid., p. 68.

vii   J. C. Plumpe, 'Vivum Saxum, Vivi Lapides: The Concept of "Living Stone" in Classical and Christian Antiquity', *Traditio*, 1943, vol. 1, pp. 1–14 (Cambridge University Press).

viii   Fabio Barry, *Painting in Stone: Architecture and the Poetics of Marble from Antiquity to the Enlightenment* (Yale University Press, 2020), p. 63.

ix   See note x to 'Marble', p. 311.

x   Sardonyx Cameo Portrait of the Emperor Augustus, Metropolitan Museum of Art, New York (online catalogue, metmuseum.org).

xi   Sardonyx Cameo of Aurora Driving her Chariot, Metropolitan Museum of Art, New York, (online catalogue, metmuseum.org).

xii   James David Draper, *Cameo Appearances* (Metropolitan Museum of Art, 2008).

xiii   Caillois, *Writing of Stones*, 'The Image in the Stone', p. 11.

PINK ANCASTER

i   Natalie Rudd, 'Keeping Watch: Expanding the Narratives of Modern British Sculpture', in *Breaking the Mould: Sculpture by Women Since 1945* (Hayward Gallery Publishing, 2020), p. 10.

ii   Hepworth's statement in the series 'Contemporary English Sculptors', *Architectural Association Journal*, April 1930, vol. XLV, no. 518, p. 384.

iii   The West Riding is one of three historic subdivisions of Yorkshire and covers the towns of Leeds, Bradford, Huddersfield, Halifax, Sheffield and York, and an enormous area of land beyond them. The West Riding was reduced in 1974, and abolished in 1986.

iv   Quoted in Caroline Maclean, *Circles and Squares: The Hampstead Modernists* (Bloomsbury, 2020), p. 21.

v   Quoted in Eleanor Clayton, *Barbara Hepworth: Art & Life* (Thames & Hudson, 2021), p. 49.

vi   Ibid., p. 50.

vii   Maclean, *Circles and Squares*, p. 23.

viii Barbara Hepworth, *Pierced Form* (1932).

ix   Anne Middleton Wagner, *Mother Stone* (Yale University Press, 2005), p. 22.

x   Ibid., p. 44.

xi   Letter from Hepworth to Nicholson, 17 January 1935, quoted in Clayton, *Barbara Hepworth*, p. 71.

QUARTZ

i   G. F. Kunz, *Gems and Precious Stones of North America* (1892), quoted in Margaret Sax, Jane M. Walsh, Ian Freestone et al., 'The Origins of Two Purportedly Pre-Columbian Mexican Crystal Skulls', *Journal of Archaeological Science*, 1 October 2008, vol. 35, no. 10, pp. 2,751–60.

ii   Jane MacLaren Walsh and Brett Topping, *The Man Who Invented Aztec Crystal Skulls: The Adventures of Eugène Boban* (Berghahn Books, 2018).

iii   Jeremy Johns and Elise Morero, 'The Diffusion of Rock Crystal Carving Techniques in the Fātimid Mediterranean' (Khalili Research Centre, University of Oxford), paper delivered to the conference, 'Beyond the Western Mediterranean: Materials, Techniques and Artistic Production', Courtauld Institute of Art, Saturday 20 April 2013, p. 1.

iv   Sax, Walsh, Freestone et al., 'Origins of . . . Mexican Crystal Skulls', pp. 15–16.

v   Landis Ehler, 'Explorers of Katmai Country: Alphonse Pinart (1852–1911)', Katmai National Park and Terrain Alaska, published online 30 October 2014.

vi   Quoted in Eric Betz, 'The Real Story Behind Aztec Crystal Skulls', *Discover*, 28 December 2020.

vii   Walsh and Topping, *Adventures of Eugène Boban*.

viii Jerry Hopkins, 'Nudie: The World's Flashiest Country and Western Stylist', *Rolling Stone*, 28 June 1969.

RED OCHRE

i   Ruth Siddall, 'Mineral Pigments in Archaeology: Their Analysis and the Range of Available Materials', *Minerals*, 8 May 2018, published online, vol. 8, no. 5:201, pp. 1–35.

ii   'Where is the Oldest Rock Art?' Africa Rock Art Archive, Bradshaw Foundation, published online.

iii   Gemma Tarlach, 'What the Ancient Pigment Ochre Tells Us About the Human Mind', *Discover*, 16 March 2018.

iv   Brooks Hays, 'Hunter-gatherers Heated Bacteria to Produce Ochre Paint Used in Pictographs', *Science News*, 20 November 2019.

v   Siddall, 'Mineral Pigments', p. 6.

vi   Ibid., p. 5.

vii   Tarlach, 'Ancient Pigment Ochre'.

viii In 1823, William Buckland identified the skeleton as a woman because it was found with beads and ornaments. He dated it to Roman times. It has more recently been identified as the body of a young man who lived perhaps 33,000–34,000 years ago.

ix   David Lewis-Williams, *The Mind in the Cave* (Thames & Hudson, 2002), pp. 253–4.

x   Chris Gosden, *The History of Magic: From Alchemy to Witchcraft, from the Ice Age to the Present* (Viking, 2020), pp. 43–6.

xi   Lewis-Williams, *Mind in the Cave*, pp. 212–13.

STONE TECHNOLOGY

i   Primo Levi, *The Periodic Table* (Picador, 1975), 'Iron'.

ii   Nicholas Le Pan, 'All the World's Metals and Minerals in One Visualization', 1 March 2020, visualcapitalist.com

iii   Lanthanum, Cerium, Praseodymium, Neodymium, Samarium, Europium, Gadolinium, Terbium, Dysprosium, Holmium, Erbium, Thulium, Ytterbium, Lutetium and Yttrium.

iv   The largest deposits of rare earth elements are hosted by carbonatites, igneous rocks derived from carbonate-rich magmas. Bradley S. Van Gosen, Philip L. Verplanck, Keith R. Long et al., 'The Rare-earth Elements—Vital to Modern Technologies and Lifestyles', U.S. Geological Survey Fact Sheet 2013–3078.

COADE STONE

i   Jane Austen, *Persuasion* (1817).

ii   John Fowles, *The French Lieutenant's Woman* (Jonathan Cape, 1969).

iii   Caroline Stanford, 'Revisiting the Origins of Coade Stone', *Georgian Group Journal*, 2016, vol. xxiv, p. 111.

iv   Roger White, introduction to Alison Kelly, *Mrs Coade's Stone* (Self Publishing Association, 1990).

v   Stanford, 'Origins of Coade Stone', p. 108.

vi   english-heritage.org.uk/learn/ histories/women-in-history/ eleanor-coade/

vii   John Summerson, *Georgian London* (1945) (Yale University Press, 2003 edn,), p. 130.

viii   Jonathan Foyle, 'How a Sculptor Cracked the Lost Mixture of Eleanor Coade's Stone', *Financial Times*, 10 July 2015.

COAL

i   John McPhee, *Annals of the Former World* (Farrar, Straus and Giroux, 1998), p. 284.

ii   There is not space here to do justice to Annie Sprinkle's extraordinary career: she has been a sex worker, porn actress and pin-up girl, and, more recently, a pioneering sex educator and performer. She wrote her PhD thesis on Providing Educational Opportunities for Sex Workers.

iii   Annie Sprinkle and Beth Stephens with Jennie Klein, 'Green Wedding to the Earth' artist statement, *Assuming the Ecosexual Position: The Earth as Lover* (University of Minnesota Press, 2021), p. 77.

iv   The 'Black Country,' is the central area of England's West Midlands and was one of the earliest parts of the country to see intensive industrialisation. It acquired the name 'Black Country' in the 1840s when it became blanketed by a choking layer of soot and coal dust.

v   James Nasmyth, Engineer, *An Autobiography* (1883), quoted in Jeremy Deller, *All That Is Solid Melts Into Air* (Hayward Publishing, 2013), p. 12

vi   *Parliamentary Papers*, 1842, vol. XVII, p. 104.

vii   'UK Fires Up Coal Power Plant as Gas Prices Soar', BBC News, 7 September 2021.

viii   'China Generated Over Half World's Coal-Fired Power in 2020: Study', Reuters, 28 March 2021.

ix   nytimes.com/ interactive/2018/12/24/climate/ how-electricity-generation-changed-in-your-state.html

x   Figures from *Goodbye Gauley Mountain* (2013).

xi   There are thought to have been a number of earlier supercontinents, dubbed Keloland, Nuna, Rodinia and Pannotia.

xii   Robert M. Hazen, *The Story of Earth* (Penguin, 2012), p. 234.

xiii   The American Geological Institute, Glossary of Geology, quoted in McPhee, *Annals*, p. 247.

xiv   Boyce is co-author of the study 'Delayed Fungal Evolution Did Not Cause the Paleozoic Peak in Coal Production', debunking a theory of coal formation that had been in currency since the 1990s. Quoted in Ker Than, 'Stanford Scientists Discover How Pangaea Helped Make Coal', *Stanford Report*, 22 January 2016.

xv   Richard Fortey, *The Hidden Landscape* (Jonathan Cape, 1993), p. 145.

COLTAN

i   Andy Robinson, 'PlayStation 2's Chaotic Japanese Launch Was 20 Years Ago Today', *Video Games Chronicle*, 4 March 2020.

ii   Lionel Sujay Vailshery, 'Cumulative Sales of Playstation 2 by Region 2021', 11 August 2021, statista.com

iii   Edoardo Totolo, 'Coltan and Conflict in the DRC', International Relations and Security Network (published online, reliefweb.int) 11 February 2009.

iv   Artur Usanov, Marjolein de Ridder, Willem Auping and Stephanie Lingermann, 'Coltan, Congo & Conflict', Polinares Case Study, The Hague Centre for Strategic Studies, 2013, no. 21, 05, p. 44.

v   Edoardo Totolo, 'Coltan and Conflict in the DRC'.

vi   Press release: 'Security Council Condemns Illegal Exploitation of Democratic Republic of Congo's Natural Resources', United Nations Security Council, 3 May 2001.

vii   Usanov, de Ridder, Auping and Lingermann, 'Coltan, Congo & Conflict', p. 59.

viii   Joe Bavier, 'Congo War-Driven Crisis Kills 45,000 a Month: Study', Reuters, 22 January 2008.

ix   'Smugglers' Paradise: Why It's Hard for Congo's Coltan Miners to Abide by the Law', *The Economist*, 21 January 2021.

x   Tom Burgis, 'Dodd-Frank's Misadventures in the Democratic Republic of Congo', *Politico*, 10 May 2015.

xi   Oluwole Ojewale, 'Child Miners: The Dark Side of the DRC's Coltan Wealth Laws and Certification Schemes Aren't Protecting the Democratic Republic of the Congo's Most Vulnerable – A Fresh Approach Is Needed', *ISS Today*, 18 October 2021.

xii   Burgis, 'Dodd-Frank's Misadventures'.

FLINT

i   *All the Year Round: A Weekly Journal, Conducted by Charles Dickens*, 9 March 1867, vol. 17, pp. 259–264.

ii   I was introduced to the story of Flint Jack through Sean Lynch's project at the Henry Moore Institute in Leeds, commissioned as part of Yorkshire Sculpture International 2019. The catalogue for the exhibition lists just some of the institutions which today exhibit Flint Jack's work: the British Museum, London; the Higgins Bedford; Hull Museums; Hunterian Museum, London; Leeds City Museum; Manchester Museum; National Museum of Ireland, Dublin; National Museum of Scotland, Edinburgh; Pitt Rivers Museum, Oxford; Royal Albert Memorial Museum; Salisbury Museums; Yorkshire Museum, York; Ulster Museum, Belfast and the Whitby Museum. From Sean Lynch and Jorge Satorre, *The Rise and Fall of Flint Jack* (Henry Moore Foundation, 2019), p. 54.

iii   Ibid., pp. 3–5.

iv   Ibid., p. 6.

v   As with all aspects of Flint Jack's biography, his years of service with two celebrated figures in the fields of geology and paleontology should be taken with a pinch of salt, and his claim to have worked for either man has been contested; see Parry Thornton, 'Edward Simpson, or, a search for "Flint Jack"' *The Geological Curator* 7(8) (The Geological Curators Group, 2002), pp. 309–317.

vi   Simon Winchester, *The Map That Changed the World: A Tale of Rocks, Ruin and Redemption* (Penguin, 2002).

vii   Lynch and Satorre, *Rise and Fall of Flint Jack*, p. 31.

viii   *All the Year Round*, 9 March 1867, vol. 17, p. 262.

ix   While it bore the byline of Henry Smithson, proprietor of the *Messenger*, the story, as told by its subject, came from Charles Monkman, a local farmer and amateur archaeologist.

x   *Malton Messenger*, 19 January 1867, quoted in Parry Thornton, 'Edward Simpson, or a Search for "Flint Jack"', *Geological Curator*, 2008, vol. 7, no. 8, p. 309.

xi   Quoted in Lynch and Satorre, *Rise and Fall of Flint Jack*, pp. 72–9.

xii   *All the Year Round*, p. 264.

xiii   George Young, *Scriptural Geology* (1838), quoted in Roger Osborne, *The Floating Egg* (Pimlico, 1999), p. 141.

HAÜYNE

i   George F. Kunz, 'The Life and Work of Haüy', *American Mineralogist*, 1918, vol. 3, pp. 60–89.

ii   Now the Jardin des Plantes of the Muséum national d'Histoire naturelle.

iii   Kunz, 'Life and Work of Haüy'.

iv   Bernadette Bensaude Vincent, 'Teaching Chemistry in the French Revolution: Pedagogy, Materials and Politics', in Lissa L. Roberts and Simon Werrett (eds), *Compound Histories: Materials, Governance and Production, 1760–1840* (Brill, 2018), p. 253.

v   Ibid.

vi   Christine Blondel, 'Haüy et l'électricité. De la démonstration-spectacle à la diffusion d'une science newtonienne', *Revue d'Histoire des Sciences* (Armand Colin, 1997), vol. 50, no. 3, p. 270.

vii   Notable figures to have studied crystal structures before Haüy include Nicholas Steno, a Danish physician who wrote on the structure of quartz crystals in 1699, and the eighteenth-century French mineralogist Jean-Baptiste Louis Romé de l'Isle.

viii   Kunz, 'Life and Work of Haüy'.

ix   René-Just Haüy, *Traité de minéralogie* (Conseil des Mines, 1801), vol. 1, pp. 23, 24.

x   Paul E. Desautels, *The Mineral Kingdom* (Paul Hamlyn, 1969), p. 37.

xi   Kunz, 'Life and Work of Haüy'.

LODESTONE

i   Theophrastus, *On Stones*, (c. 350–287 BCE), ed. and trans. by Earle R. Caley and John F. C. Richards (Ohio State University Press, 1956), § 5.

ii   Claudian, *Magnes*, ll. 43–4.

iii   D. W. Emerson, 'The Lodestone, from Plato to Kircher', *Preview*, 2014, no. 173, pp. 52–62.

iv   Patti Wigington 'What is Lodestone?' Learn Religions, 4 April 2019 (published online, learnreligions.com).

v   Susan Silverman, 'Compass, China, 220 BCE', Museum of Ancient Inventions, Smith College History of Science (published online, smith. edu).

vi   'The Chinese Compass and the Birth of Navigation', Sichuan Museum, Chengdu, Google Arts & Culture (published online, artsandculture.google.com).

vii   Amir D. Aczel, *The Riddle of the Compass: The Invention that Changed the World* (Harcourt, 2001).

viii   Henrich Winter, 'Who Invented The Compass?' *Mariner's Mirror*, 1937, vol. 23, no. 1, pp. 95–102.

MICA

i   Ilana Halperin, *Minerals of New York* (map), 2017.

ii   Richard Rogers and Oscar Hammerstein II, 'The Surrey with The Fringe on Top', *Oklahoma!* (1943).

iii   Arthur Bevan, 'War Minerals: Mica', *Scientific Monthly*, vol. 60, no. 5 (American Association for the Advancement of Science, 1945), p. 393.

iv   'Mica quarrying and processing in Scotland during the Second World War', *Earthwise*, 9 January 2018, from: R. P. McIntosh, 'Mica Quarrying and Processing in Scotland during the Second World War', *Journal of the Russell Society*, 2001, vol. 7, Part 2, pp. 71–4.

v   Quoted in Barbara Müller-Wesemann, 'Martha Glass, "Every day in Theresin is a gift"', Key Documents of German-Jewish History: A Digital Source Edition (published online, jewish-history-online.net).

vi   Jaap W. Focke, '1943 to 1946: Survivors', *Machseh Lajesoumim: A Jewish Orphanage in the City of Leiden, 1890–1943* (Amsterdam University Press, 2021).

vii   33,000 Jews perished at Theresienstadt – almost one in four – 87,000 were deported to the Auschwitz camps (Auschwitz-Birkenau State Museum).

viii   Production facilities in the Theresienstadt ghetto (published online, ghetto-theresienstadt.de).

ix   Robert M. Ehrenreich and Jane Klinger, 'For A Piece Of Mica: A Holocaust Survival Story', The Ultimate History Project blog.

MILLSTONE GRIT

i   Richard Fortey, *The Hidden Landscape* (Jonathan Cape, 1993), p. 153.

ii   Russell H. Anderson, 'The Technical Ancestry of Grain-Milling Devices', *Agricultural History*, July 1938 (Agricultural History Society), vol. 12, no. 3, p. 256.

iii   Ian Kuijt and Bill Finlayson, 'Evidence for Food Storage and Predomestication Granaries 11,000 Years Ago in the Jordan Valley', *PNAS*, 7 July 2009, vol. 106 , no. 27, pp. 10,966–70.

iv   L. A. Moritz, 'Vitruvius' Water-Mill', *Classical Review*, December 1956, vol. 6, nos 3/4 (Cambridge University Press on behalf of The Classical Association), pp. 193–6.

v   Sophia Germanidou, 'Watermills in Byzantine Greece (Fifth–Twelfth Centuries): A Preliminary Approach to the Archaeology of Byzantine Hydraulic Milling Technology', *Byzantion*, 2014, vol. 84 (Peeters Publishers), pp. 185–201.

vi   Currer Bell (Charlotte Brontë), preface to 1850 edition of Emily Brontë's *Wuthering Heights* (1847).

vii   Anne Carson, 'The Glass Essay' from *Glass, Irony, and God* (New Directions Publishing Corporation, 1995).

OBSIDIAN

i   Amy L. Covell-Murthy, 'Minecraft™ Lied To Me!?' Carnegie Museum of Natural History blog.

ii   Miyasaka Kiyoshi, 'Hoshigatō Obsidian Source Historic Site', Japanese Archaeological Association blog, 2016.

iii   'Surgeons use Stone Age Technology for Delicate Surgery', *University Record*, 10 September 1997 (University of Michigan, 1997).

iv   Sherry Yi and H. Chad Lane, 'Playing with Virtual Blocks: Minecraft as a Learning Environment for Practice and Research', *Cognitive Development in Digital Contexts*, December 2017, pp. 145–66.

v   Clive Thompson, 'The Minecraft Generation: How a Clunky Swedish Computer Game is Teaching Millions of Children to Master the Digital World', *New York Times*, 14 April 2016.

vi   Hana Schank, 'The Myth of the Minecraft Curriculum', *Atlantic*, 20 February 2015.

vii   J. Clement, 'Minecraft Active Player Count Worldwide 2016–2021', *Statista*, 9 December 2021, statista.com

viii   A. L. Kroeber, *Handbook of the Indians of California* (California State University, 1925), pp. 197–8, quoted in Susan Fox Hodgson, 'Obsidian: Sacred Glass from the California Sky,' in L. Piccardi and W. B. Masse (eds), *Myth and Geology*, 1 January 2007 (Geological Society), Special Publications, vol. 273, p. 279.

ix   Hodgson, 'Obsidian'.

x   Ibid., p. 298.

LIVING
STONES

i   Louise Firth, 'Artificial Islands: "Leviathan" Hybrids of Nature and Technology', n. d., Leviathan-cycle. com

ii   Robert M. Hazen, D. Papineau, W. Bleeker et al., 'Mineral Evolution', *American Mineralogist*, 2008, vol. 93, pp. 1,693–720.

iii   Jolyon Ralph, 'Mineral Evolution', n. d., mindat.org

BLUE
LIAS

i   Richard Fortey, *The Hidden Landscape* (Jonathan Cape, 1993), p. 177.

ii   Christopher McGowan, *The Dragon Seekers* (Little, Brown, 2002), p. 13.

iii   The formation also appears in the north east of England, near Whitby.

iv   Augustine, *City of God*, Book XV, Ch. 9 (written c. 426).

v   Nigel J. Clarke, *Mary Anning 1799–1847: A Brief History* (Nigel J. Clarke Publications, 1998), p. 4.

vi   Tom Sharpe, *The Fossil Woman* (Dovecot Press, 2020), p. 70.

vii   McGowan, *Dragon Seekers*, p. 22.

viii   Sharpe, *Fossil Woman*, passim.

ix   Anning was so devoted to Tray that he seems to make a posthumous appearance in her portrait, credited to William Grey, and painted in February 1842.

x   Sharpe, *Fossil Woman*, p. 41.

xi   Ibid., passim.

xii   Quoted in Ibid., p. 115.

xiii   Quoted in Ibid., pp. 145–6.

xiv   McGowan, *Dragon Seekers*, p. 216.

CALCULI

i   Four of Lonsdale's siblings had died in early infancy.

ii   As just one example, Lonsdale's work in the 1920s 'showed conclusively that the benzene ring was flat, something that chemists had been arguing about for 60 years.' An important milestone in organic chemistry. Dr Peter Childs, 'Woman of Substance', *Chemistry in Britain*, January 2003, vol. 39, no. 1, pp. 41–3.

iii   2 per cent of minerals are named after women, and of those, few are named after women scientifically involved in the field. 'Out of the 5493 unique minerals currently recognized by the International Mineralogical Association, only 112 minerals are named after women. Ninety-six different women have one or more minerals named after them; however, many of those women were honored either as a gift from a man (despite not being involved in geology) or with their husbands.' Sophia Brooks-Randall, 'Womenclature', blog of the Mineralogical & Geological Museum, University of Harvard, 2019, mgmh.fas.harvard.edu/ womenclature

iv   Dorothy M. C. Hodgkin, O.M., F.R.S, 'Kathleen Lonsdale', *Biographical Memoirs of Fellows of the Royal Society*, vol. 21 (Royal Society Publishing, 1974), p. 473.

v   Professor Dame Kathleen Lonsdale, *Woman's Hour*, BBC Radio Two, first broadcast 14:00, Wednesday 8 November 1967.

vi   Hodgkin, 'Kathleen Lonsdale', p. 452.

vii   Following a manifesto drawn up by Bertrand Russell and Albert Einstein in 1955, the Pugwash Conferences on Science and World Affairs aimed 'to bring scientific insight and reason to bear on . . . the catastrophic threat posed to humanity by nuclear and other weapons of mass destruction.' Lonsdale was an advisor to co-founder Sir Joseph Rotblat, awarded the 1995 Nobel Peace Prize, pugwash.org

viii   Hodgkin, 'Kathleen Lonsdale', p. 453.

ix   Ibid., p. 472.

x   Kathleen Lonsdale, 'Human Stones', *Science*, 15 March 1968, New Series, vol. 159, no. 3, 820, pp. 1,199–207, American Association for the Advancement of Science.

xi   Helen E. Maynard-Casely, 'Franklin and Lonsdale: Two Role Models For Our Time', *Canadian Journal of Physics*, 26 September 2017.

COPROLITE

i   Both *A Coprolitic Vision* and Buckland's coprolite tabletop are pictured among the plates in Tom Sharpe's *The Fossil Woman: A Life of Mary Anning* (Dovecot Press, 2020).

ii   The future King George IV, who served as Prince Regent from 1811–20.

iii   Quoted in Christopher McGowan, *The Dragon Seekers* (Little, Brown, 2001), p. 36.

iv   'The Heart of a King', Geological Society of London blog, 23 December 2014.

v   Henry Acland, quoted in McGowan, *Dragon Seekers*, p. 33.

vi   William Buckland, *Reliquiae Diluvianae, Or, Observations on the Organic Remains Contained in Caves, Fissures, and Diluvial Gravel, and on Other Geological Phenomena, Attesting the Action of an Universal Deluge* (John Murray, 1824), p. 11.

vii   Ibid., p. 12.

viii   The Exeter 'Change (or more properly, Exeter Exchange) was a superior shopping arcade on the Strand in London: from 1773 the top floor housed a menagerie of exotic animals and birds.

ix   McGowan, *Dragon Seekers*, p. 84.

x   Ibid., p. 190.

CORAL

i   Corals tutorial, National Ocean Service, (published online, oceanservice.noaa.gov) National Oceanic and Atmospheric Administration, US Department of Commerce.

ii   Donna Haraway, *Staying With the Trouble: Making Kin in the Chthulucene* (Duke University Press, 2016), p. 80.

iii   'At What Price? The Economic, Social And Icon Value Of The Great Barrier Reef', Deloitte Access Economics for the Great Barrier Reef Foundation (published online, deloitte.com, 2017).

iv   This telling and interpretation is based on verbal accounts by Gudju Gudju Fourmile for Australian Earth Laws Alliance (2015), and for the Gondwana Indigenous Choir (2017).

v   Traditional Owner Groups are Aboriginal people of local descent who have particular rights and responsibilities in relation to a piece of land or an area of the sea.

vi   Nick Reid and Patrick D. Nunn, 'Ancient Aboriginal Stories Preserve History of a Rise in Sea Level', *Conversation*, 12 January 2015.

vii   T. Hughes,, J. Kerry, M. Álvarez-Noriega et al., 'Global Warming and Recurrent Mass Bleaching of Corals', *Nature*, 2017, vol. 543 , pp. 373–7.

viii   Gerry Turpin, a Mbarbaram man from northern Queensland and ethnobotanist at the Queensland Herbarium, quoted in Emilie Ens and Alana Grech, 'Indigenous Ranger Programs Are Working in Queensland – They Should Be Expanded', *Conversation*, 14 January 2018.

ix   Jaume Rius Lopez, 'Australia: As the Great Barrier Reef Shrinks, Torres Strait Islanders Have Everything to Lose', 2019, minorityrights.org

x   citizensgbr.org

xi   Curator's Comments on 'Corslet made from a mesh of coral beads on vegetable fibre', item Af1944,04.63 in the online catalogue of the British Museum.

xii   Ovid, *Metamorphoses* (c. 8 CE), Book 4, Part 9, l. 740.

xiii   Kiki Karoglou, 'Dangerous Beauty: Medusa in Classical Art', *Metropolitan Museum of Art Bulletin*, winter 2017, vol. 75, no. 3, p. 32.

LEWISIAN GNEISS

i   Richard Fortey, *The Hidden Landscape* (Jonathan Cape, 1993), p. 32.

ii   Boris Robert, Mathew Domeier and Johannes Jakob, 'On the Origins of the Iapetus Ocean', Earth-Science Reviews, October 2021, vol. 221, published online 4 September 2021.

iii   Henry Faul and Carol Faul, *It Began With a Stone: A History of Geology from the Stone Age to the Age of Plate Tectonics* (Wiley Interscience, 1983), pp. 225–30.

iv   Ibid., p. 229.

v   Fortey, *Hidden Landscape*, p. 43.

vi   Formerly Callanish.

vii   Aubrey Burl, *From Carnac to Callanish: The Prehistoric Stone Rows and Avenues of Britain, Ireland, and Brittany* (Yale University Press, 1993), p. 61.

viii   Ibid., p. 64.

ix   Ibid., p. 11.

PEARL

i   'The Hymn of the Pearl' ('The Hymn of Judas Thomas the Apostle in the Country of the Indians'), trans. by G. R. S. Mead (1908), § 3, 21.

ii   Isidore of Seville, *Etymologies* (c. 615–30), Book XII.vi.49.

iii   There has been considerable confusion over the identity of this pearl, which was long thought to be La Peregrina, bought by Richard Burton for Elizabeth Taylor in 1969. The current historical mood suggests this is not the case, but that may be because a very persuasive jewellery company put a vast pearl purported to be the Mary Tudor gem on the market in 2013. A rather convincing piece of historical sleuthing can be found in the portrait analysis on EraGem's blog: 'A Tale of Two Pearls: Tracing La Peregrina & Mary Tudor's Pearl Through Portraits', 28 December 2013. Nevertheless, it may all be marketing hooey.

iv   Victoria Finlay, *Jewels: A Secret History* (Sceptre, 2005), p. 111.

v   Quoted in Cassandra Auble, 'The Cultural Significance of Precious Stones in Early Modern England', thesis (published online, University of Nebraska-Lincoln, 2011), p. 34.

vi   Ibid., p. 31.

vii   Finlay, *Jewels*, p. 102.

viii   Wilburn Dowell Cobb, 'The Pearl of Allah', *Natural History*, November 1939.

ix   For the true story of Cobb, Barbish and the Pearl of Allah, I am much indebted to the investigative journalism of Michael LaPointe in his 'Chasing The Pearl Of Lao Tzu', *Atlantic*, June 2018.

SLATE

i   I am indebted to John McPhee for this observation.

ii   F. J. North, Bruce Campbell and Richenda Scott, *Snowdonia* (Collins, 1949), p. 74.

iii   Ibid., pp. 75–6.

iv   'Manod: The Nation's Treasure Caves', February 2018, National Gallery, London (published online, natonalgallery.org).

v   'Story of Slate', National Slate Museum, Llanberis, Gwynedd.

vi   During the 1830s Adam Sedgwick named the Cambrian, his rival Roderick Murchison the Silurian formation. In 1879 Charles Lapworth identified the intervening Ordovician: John Rickus, 'Rocks and Geology: The Welsh Connection', 2013, *Transactions of the Honourable Society of Cymmrodorion*, New Series, vol. 19, pp. 169–74.

vii   Jan Zalasiewicz, *The Planet in a Pebble* (Oxford University Press, 2012), p. 62.

viii   Ibid., p. 154.

ix   Wirt Sikes, *British Goblins: Welsh Folk-lore, Fairy Mythology, Legends and Traditions* (James R. Osgood & Co., 1881), p. 27.

SULPHUR

i   Sir William Hamilton, *Campi Phlegraei* (1776), vol. 2.

ii   From the letters of Sir William Hamilton, quoted in Geoffrey Stone, 'Sir William Hamilton, Networks and Knowledge' (Department of Humanities University of Roehampton, 2020).

iii   Revelation xxi, 8, King James version.

iv   *Libellus de Alchimia, Ascribed to Albertus Magnus*, ed. and trans. by Sister Virginia Heines (1958), from Stanton J. Linden (ed.), *The Alchemy Reader: From Hermes Trismegistus to Isaac Newton* (Cambridge University Press, 2003), p. 101.

v   Aristotle, *Meteorology* (c. 340 BCE), trans. E. W. Webster (Princeton University Press, 1984).

vi   The 'Islamic Golden Age' or 'Golden Age of Islam' are terms used for the period between roughly 622 and 1250 CE.

vii   Linden (ed.), *Alchemy Reader*, p. 97.

viii   Ibid., p. 5.

ix   Nathan Schwartz-Salant, *C. G. Jung on Alchemy* (Routledge, 1995), 'Introduction', p. 11.

x   Ibid., p. 4.

xi   Interestingly, the two elements are assigned differently in the Indian tradition of alchemy – see 'Cinnabar', p. 78

xii   Quoted in Linden (ed.), *Alchemy Reader*, p. 115.

xiii   'The Personifications of the Opposites' (1955–6), quoted in Schwartz-Salant, *C.G. Jung on Alchemy*, p. 71.

THE UJARAALUK UNIT

i   Robert M. Hazen, *The Story of Earth* (Penguin, 2012).

ii   Ibid.

iii   Jonathan O'Neil, Don Francis and Richard W. Carlson, 'Implications of the Nuvvuagittuq Greenstone Belt for the Formation of Earth's Early Crust', *Journal of Petrology*, May 2011, vol. 52, no. 5, pp. 985–1,009.

iv   Richard W. Carlson, Marion Garçon, Jonathan O'Neil et al., 'The Nature of Earth's First Crust', *Chemical Geology*, 2019, vol. 530, published online 15 October 2019.

v   Timothy Morton, *Hyperobjects: Philosophy and Ecology after the End of the World* (University of Minnesota Press, 2013), p. 60.

vi   John McPhee, *Annals of a Former World* (Farrar Straus and Giroux, 1998), p. 90.

vii   Hazen, *Earth*, p. 30.

ACKNOWLEDGEMENTS

I AM INDEBTED FIRST OF ALL TO GEORGINA LAYCOCK OF JOHN Murray who brought this book to me as an idea and must take full credit for the concept. Also to Abigail Scruby, who endured the editing process with good humour, and has infected me with the term 'signposting'.

I have benefitted enormously from the generosity of friends and colleagues who have shared research, recommended texts and sustained me with their enthusiasm. Heartfelt thanks to: Didier Bouakaze-Khan for, among many other things, sharing his excitement about the work of David Lewis-Williams. Pio Abad and Frances Wadsworth Jones for discussing their research into Imelda Marcos's jewellery collection. Joanna Hardy for directing me to Jack Ogden's essay on spinel, and Robert Violette for sending me copies of Joanna's books *Ruby*, *Emerald* and *Sapphire*. Laima Leyton for alerting me to Ametista do Sul. Annie Sprinkle and Beth Stephens for allowing me to share the story of their marriage to the Appalachian Mountains. Sissel Tolaas for telling me about her extraordinary research project at Pompeii. Lily Jencks for giving me access to specialist volumes in her parents' library for Lingbi. Cristiano Ferraris at the Muséum national d'Histoire naturelle, Paris, for helping with my enquiries regarding René Just Haüy. Sophie Williamson for sending me Ursula K. Le Guin's text 'On Being Taken for Granite'. Victoria Mills for sharing her essay 'Dandyism, Visuality and the "Camp Gem"'. Paul Devereux and Jon Wozencroft for sharing their research on the acoustic properties of dolerite and the landscape of South Wales, 'Stone Age Eyes and Ears'.

Elisa Vazquez de Gey for sharing excerpts from the writing of Anita Delgado. Kieran McCarthy of Wartski for talking about Carl Fabergé's love of aquamarines. Ilana Halperin for her generous book recommendations, and for allowing me to tell a little of her mica story. Matt Kirsch of the Noguchi Museum for sending links, pointers and his essay, 'The Search for Stones' (and thanks to Florence Ostende for connecting us). Lindsay Allason-Jones for posting me a photocopy of her excellent book on Roman jet. Sarah Victoria Turner for lending me her copy of *Mother Stone*. Rebecca Mead for sending me a scan of Rodney Castleden's *The Cerne Giant*. Emma Cousin for sharing the sucking stone sequence from Samuel Beckett's *Molloy*. Meli McBurnie for her endless enthusiasm, and for introducing me to Wally Brown's stories and the *Mani* stones of Tibetan Buddhism. Alice Hertzog for contributing to my *Sisyphusarbeit* with research on the stone axe heads of Nigeria (and with apologies for not including the *Unspunnenstein*.)

Elaine and Peter, thank you for letting me stay in your house in Chite during the early stages of research on this book – the rocky view across the valley to the high sierra was powerful inspiration. Flan and Paul, thank you for lending me your London home so that I had space and quiet to write over Christmas 2021.

Profound thanks to Dr Maria Souvatzi and Dr Emma V. Carrington FRCS, to their teams and to the staff at St Mary's Paddington and the Charing Cross hospitals. Without them I would not have been here to write this book.

I am grateful to Ben and Isaac for being my first (and best) readers, and who braved the dragon under the mountain to offer critical feedback. And Ben (again, and always) for clambering over many stones with me, actual and metaphorical.

INDEX

References to images are in *italics*.